AF595035

Advancement in the Fluid Dynamics Research of Reversible Pump-Turbine

Advancement in the Fluid Dynamics Research of Reversible Pump-Turbine

Editors

Ran Tao

Changliang Ye

Xijie Song

MDPI • Basel • Beijing • Wuhan • Barcelona • Belgrade • Manchester • Tokyo • Cluj • Tianjin

Editors
Ran Tao
College of Water Resources and Civil Engineering
China Agricultural University
Beijing
China

Changliang Ye
College of Energy and Electrical Engineering
Hohai University
Nanjing
China

Xijie Song
Department of Energy and Power Engineering
Tsinghua University
Beijing
China

Editorial Office
MDPI
St. Alban-Anlage 66
4052 Basel, Switzerland

This is a reprint of articles from the Special Issue published online in the open access journal *Water* (ISSN 2073-4441) (available at: www.mdpi.com/journal/water/special_issues/Fluid_Reversible_Pump_Turbine).

For citation purposes, cite each article independently as indicated on the article page online and as indicated below:

LastName, A.A.; LastName, B.B.; LastName, C.C. Article Title. *Journal Name* **Year**, *Volume Number*, Page Range.

ISBN 978-3-0365-5858-5 (Hbk)
ISBN 978-3-0365-5857-8 (PDF)

© 2022 by the authors. Articles in this book are Open Access and distributed under the Creative Commons Attribution (CC BY) license, which allows users to download, copy and build upon published articles, as long as the author and publisher are properly credited, which ensures maximum dissemination and a wider impact of our publications.
The book as a whole is distributed by MDPI under the terms and conditions of the Creative Commons license CC BY-NC-ND.

Contents

Editorial

Pumped Storage Technology, Reversible Pump Turbines and Their Importance in Power Grids

Ran Tao [1,2,*], Xijie Song [3] and Changliang Ye [4]

1 College of Water Resources and Civil Engineering, China Agricultural University, Beijing 100083, China
2 Beijing Engineering Research Center of Safety and Energy Saving Technology for Water Supply Network System, China Agricultural University, Beijing 100083, China
3 State Key Laboratory of Hydroscience and Engineering & Department of Energy and Power Engineering, Tsinghua University, Beijing 100084, China
4 College of Energy and Electrical Engineering, Hohai University, Nanjing 211100, China
* Correspondence: randytao@cau.edu.cn

Citation: Tao, R.; Song, X.; Ye, C. Pumped Storage Technology, Reversible Pump Turbines and Their Importance in Power Grids. *Water* **2022**, *14*, 3569. https://doi.org/10.3390/w14213569

Received: 20 October 2022
Accepted: 23 October 2022
Published: 6 November 2022

Publisher's Note: MDPI stays neutral with regard to jurisdictional claims in published maps and institutional affiliations.

Copyright: © 2022 by the authors. Licensee MDPI, Basel, Switzerland. This article is an open access article distributed under the terms and conditions of the Creative Commons Attribution (CC BY) license (https://creativecommons.org/licenses/by/4.0/).

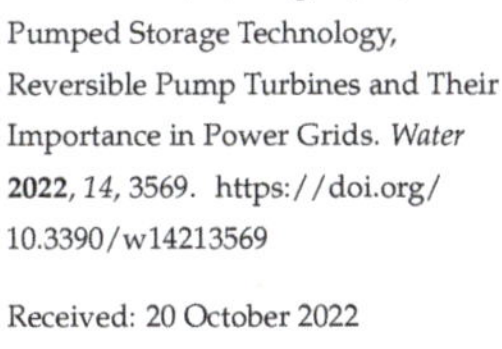

1. The Pumped Storage System and Its Constituent Elements

Pumped storage hydro is a mature energy storage method. It uses the characteristics of the gravitational potential energy of water for easy energy storage, with a large energy storage scale, fast adjustment speed, flexible operation and high efficiency [1]. The pumped storage power station, as the equipment for the peak shaving, frequency modulation and phase modulation of the power grid, has been applied in recent decades and can effectively compensate for the instability of the power grid. As shown in Figure 1, in order to store energy in the form of the mechanical energy of water, an upper reservoir and a lower reservoir are necessary. Penstock is used to connect the two reservoirs. The key components of a pumped storage power station are the hydro turbine and pump, which usually adopt the form of bladed hydraulic machinery. The mechanical energy of the water and the mechanical energy of the runner can be converted to each other. The mechanical energy of the runner depends on the mutual interaction between the generator, or motor, and the electrical energy. In recent years, because of a series of significant advantages, the runners and motors of pumped storage units have come to be designed as reversible [2,3]. At the peak level of power consumption during the day, water flows from the lower reservoir into the reservoir. The mechanical energy of the water is converted into the mechanical energy of the runner and then into electrical energy in order to generate electricity. When the power consumption is low at night, the motor drives the runner to rotate, pumping water from the lower reservoir into the upper reservoir for its storage. Pumped storage technology is simple in principle, powerful in function and significant in terms of engineering [4].

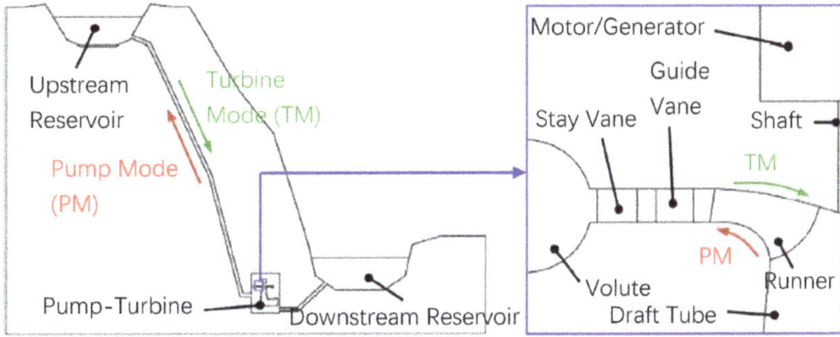

Figure 1. Schematic Map of the Pumped Storage Hydro Unit.

2. Reversibility of Bladed Hydraulic Machinery

Bladed hydraulic machinery is the foundation of energy conversion in pumped storage technology. According to the theorem of the moment of momentum, the change in the moment of momentum of an axis in unit time is equal to the sum of all the external forces acting on the control body on the same axis. When the external torque is 0, the moment of momentum of the control body remains unchanged. The bladed hydraulic machinery is regarded as the control body (see Figure 2), and the change in the moment of momentum ΔL of the fluid can be written as:

$$\Delta L = \rho Q(r_1 C_{u1} - r_2 C_{u2}) \tag{1}$$

where ρ is the density, Q is the volumetric flow rate, r is the radius position of the blade, C_u is the circumferential component of the absolute velocity C, and subscripts 1 and 2 represent the runner inlet and outlet. In an ideal state, the energy lost by the water flow is equal to that obtained by the runner:

$$\rho QgH = M\omega \tag{2}$$

where ω is the rotational angular speed of the runner, and H is the energy difference between the runner inlet and outlet. According to the theorem of the moment of momentum, the change in the moment M and moment of momentum ΔL is equal:

$$\rho Q(r_1 C_{u1} - r_2 C_{u2}) = \rho QgH/\omega \tag{3}$$

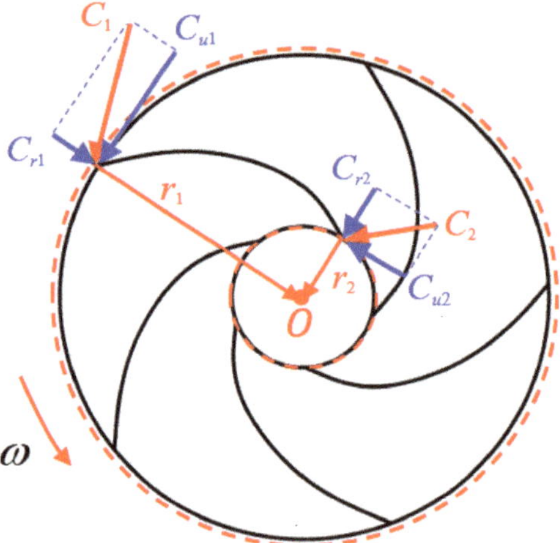

Figure 2. Change in the velocity and moment of momentum at the inlet and outlet of the runner control body.

If U represents the linear speed of the runner, we obtain:

$$U = r\omega \tag{4}$$

$$(U_1 C_{u1} - U_2 C_{u2})/g = H \tag{5}$$

As for the relationship between U and C_u and the relationships between the other velocity components, the velocity triangle shown in Figure 3 is commonly used in bladed hydraulic machinery [5]. The relative speed W is controlled by the blade shape, and the axial speed C_m is affected by the flow rate. At a certain flow rate, the reasonable design of the blades can cause the runner to convert the ideal energy H. For a hydro turbine runner whose motor drives it to rotate reversely (the rotational speed is unchanged), the U magnitude is the same, and the direction is the opposite. The blade shape provides the fluid with the same relative velocity W in the opposite direction. The resultant absolute velocity C is the same in terms of the magnitude and opposite in terms of the direction. Components such as C_u and C_m also have the same magnitude and reverse direction. According to Equation (5), the fluid obtains energy H and flows to the upstream reservoir, and the turbomachinery is switched to the pump mode. According to Equations (1) and (2), the momentum moment increases after the fluid flows in and out of the runner. Under ideal conditions, the energy lost by the runner is equal to the energy obtained by the water flow. There are losses under non-ideal conditions, and the runner efficiency in the turbine mode η_t and pump mode η_p can be calculated by the following equations:

$$\eta_t = M\omega/\rho QgH \quad (6)$$

$$\eta_p = \rho QgH/M\omega \quad (7)$$

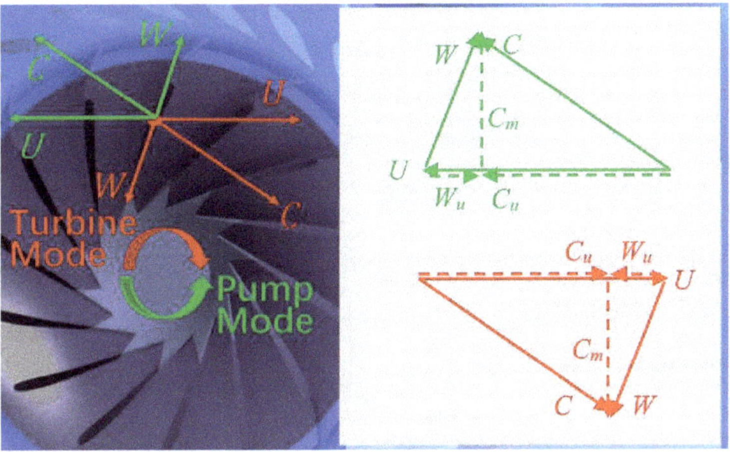

Figure 3. Velocity triangles of the bladed hydraulic machinery's runner in the turbine mode and pump mode.

The reversibility of the bladed hydraulic machinery provides a good solution for the problem of pumped storage technology. One design can be applied to two working modes. The reverse rotation of the runner completes the conversion of the working modes and responds to the demands of the power grid. From an economic perspective, the pumped storage is generally designed with a head of up to 800 m, and the runner is the Francis type [6]. For the tidal energy, the rising and ebbing tides can be dammed in the bay to realize the forward and reverse pumping and forward and reverse power generation. The low-head H, full-condition pumped storage hydro units become feasible, similar to the tubular turbine or axial flow pump [7]. It can also achieve a two-way efficient operation.

3. Cooperation between Pumped Storage and Renewable Energy

China strives to reach peak carbon dioxide emissions by 2030 and achieve the goal of carbon neutrality by 2060. Reducing the use of fossil energy and increasing the proportion of renewable energy are important goals. At present, with the rapid growth of wind power generation and solar power generation, there is a serious problem of instability. Many renewable energy sources, including wind energy, solar energy, tidal energy, wave energy and ocean current energy, require the cooperation with respect to large-scale energy storage technology [8]. In order to ensure the security and stability of the power system, many countries have built a large number of pumped storage power plants to regulate energy flexibly, efficiently and cleanly. In many developed countries, the proportion of pumped storage power plants in the power system exceeds 10%. At present, the global installed capacity of pumped storage exceeds 160 million kW, accounting for more than 94% of the total energy storage capacity. More than 100 pumped storage projects are under construction, which aim to realize the cooperation with renewable energy demands. High-head, large-capacity, and variable-speed pumped storage units are the focus of subsequent development and construction. The study of the flow problems of vane-type hydraulic machinery pumps and turbines is of great significance for the stable operation of pumped storage units. In particular, in the development of pumped storage technology, this is very important for efforts to clarify the *Q-H* stability characteristics, start-up *S*-characteristics, inception cavitation, pressure pulsation and other issues.

Author Contributions: Writing—original draft preparation, R.T., X.S. and C.Y.; writing—review and editing, R.T., X.S. and C.Y.; All authors have read and agreed to the published version of the manuscript.

Funding: This research received no external funding.

Acknowledgments: We would like to acknowledge the communication and cooperation of our colleagues from China Agricultural University, Tsinghua University, Hohai University, the State Key Laboratory of Hydro-Power Equipment (China), State Key Laboratory of Hydroscience and Engineering (China), State Grid Xinyuan Company, CSG Power Generation, DEC, HEC, and ANDRITZ Hydro. We also acknowledge the support of the National Natural Science Foundation of China.

Conflicts of Interest: The authors declare no conflict of interest.

References

1. Mei, Z. *Technology of Pumped-Storage Power Generation*; China Machine Press: Beijing, China, 2000.
2. Tanaka, H.; Tsunoda, S. The development of high head single stage pump turbines. In Proceedings of the 10th IAHR Symposium on Hydraulic Machinery and Cavitation, Tokyo, Japan, 28 September–2 October 1980.
3. Oishi, A.; Yokoyama, T. Development of high head single and double stage reversible pump-turbines. In Proceedings of the 10th IAHR Symposium on Hydraulic Machinery and Cavitation, Tokyo, Japan, 28 September–2 October 1980.
4. Tao, R.; Zhou, X.; Xu, B.; Wang, Z. Numerical investigation of the flow regime and cavitation in the vanes of reversible pump-turbine during pump mode's starting up. *Renew. Energy* **2019**, *141*, 9–19. [CrossRef]
5. Gulich, J.F. *Centrifugal Pumps*, 2nd ed.; Springer: Berlin/Heidelberg, Germany, 2010.
6. Liu, D. Study on Key Technologies of Variable Speed Pumped Storage Units and Pumps in China. *Hydropower Pumped Storage* **2020**, *32*, 2–3.
7. Ahn, S.H.; Zhou, X.; He, L.; Luo, Y.; Wang, Z. Numerical estimation of prototype hydraulic efficiency in a low head power station based on gross head conditions. *Renew. Energy* **2020**, *153*, 175–181. [CrossRef]
8. Bhimaraju, A.; Mahesh, A.; Joshi, S.N. Techno-economic optimization of grid-connected solar-wind-pumped storage hybrid energy system using improved search space reduction algorithm. *J. Energy Storage* **2022**, *52*, 104778. [CrossRef]

MDPI

Article

Numerical Simulation of Internal Flow Characteristics and Pressure Fluctuation in Deceleration Process of Bulb Tubular Pump

Jiaxu Li [1], Fengyang Xu [2], Li Cheng [1,*], Weifeng Pan [3], Jiali Zhang [4], Jiantao Shen [1] and Yi Ge [4]

1 College of Hydraulic Science and Engineering, Yangzhou University, Yangzhou 214000, China; lijiaxu_yzu@163.com (J.L.); shenjiantao888@163.com (J.S.)
2 Jiangsu Zhenjiang Jianbi Pumping Station Management Office, Zhenjiang 212006, China; xyliu98@126.com
3 Luoyun Water Conservancy Project Management Divison in Jiangsu Province, Suqian 223800, China; jssqpwf@163.com
4 Jurong Water Conservancy Bureau in Jiangsu Province, Jurong 212499, China; jrfxfh@126.com (J.Z.); geyi1988@126.com (Y.G.)
* Correspondence: chengli@yzu.edu.cn

Abstract: In order to explore the change in internal and external characteristics and the pressure fluctuation of the large bulb tubular pump unit during deceleration, a transient and steady three-dimensional (3D) numerical simulation is executed, based on the standard k-ε turbulence model and the change in boundary conditions such as flow rate. Finally, the pressure fluctuation data are analyzed by the wavelet method. There is a good agreement between the experimental data and numerical simulation results. During the deceleration process of the unit, the head decreases linearly while the efficiency remains stable. Meanwhile, the shock phenomenon and hysteresis effect appear before and after the unit head deceleration. Although there are vortex and backflow in the outlet conduit during deceleration, the pressure distribution on the suction surface of the impeller blades changes uniformly and significantly. The pressure fluctuation changes on the inlet surface of the impeller are more obvious during the deceleration: the closer to the hub, the greater the pressure, and this change decreases with decreasing radius. The fluctuation energy is mainly concentrated in the high-frequency region of 100–120 Hz and decreases uniformly with the deceleration of the rotational speed. This paper provides a reference for the energy utilization and safe operation of the water pump unit in adjusting speeds with variable frequency.

Keywords: bulb tubular pump; numerical simulation; adjusting speed; transition process; pressure fluctuation

Citation: Li, J.; Xu, F.; Cheng, L.; Pan, W.; Zhang, J.; Shen, J.; Ge, Y. Numerical Simulation of Internal Flow Characteristics and Pressure Fluctuation in Deceleration Process of Bulb Tubular Pump. *Water* **2022**, *14*, 1788. https://doi.org/10.3390/w14111788

Academic Editors: Changliang Ye, Xijie Song and Ran Tao

Received: 19 April 2022
Accepted: 31 May 2022
Published: 2 June 2022

Publisher's Note: MDPI stays neutral with regard to jurisdictional claims in published maps and institutional affiliations.

Copyright: © 2022 by the authors. Licensee MDPI, Basel, Switzerland. This article is an open access article distributed under the terms and conditions of the Creative Commons Attribution (CC BY) license (https://creativecommons.org/licenses/by/4.0/).

1. Introduction

A pump that can convert mechanical energy into liquid kinetic energy and convey fluid directionally has been widely used in many fields [1]. Among the different types of pumps, the tubular pump is widely used in low-head pumping stations due to its several advantages, such as high efficiency, good hydraulic performance, and compact structure [2]. Compared with the axial-flow pump and the mixed-flow pump, the use of the tubular pump unit can reduce the amount of excavation of the factory and the concrete needed, and then greatly reduce the overall cost of the pumping station [3,4].

Although the hydraulic performance of the tubular pump is stable under the designed operating condition, the operation point of the pump will inevitably alter due to the change in the internal and external factors. In order to ensure the efficient and reliable operation of the pump to reduce energy consumption and mechanical losses, adjusting the speed with Variable-Frequency Drives (VFDs) is one of the effective and feasible methods. As part of the Eastern Route of South-to-North Water Diversion Project, many large-scale tubular pumping stations have made use of adjusting the speed with VFDs to regulate the operation point to achieve the best efficiency point [5].

The one-dimensional (1D) flow model is an effective method that balances computational accuracy and speed, which is used to analyze flow characteristics and pressure distribution normally. Gu et al. [6,7] established a new self-closed 1D pressure model by introducing Poncet's K formula. Based on the above theory, the pressure of the casing wall can be predicted well. They also proved that the rotating speed is related to thrust acting on the shrouds, which has little effect on volumetric efficiency. Taking into consideration the blade slip factor, the self-closed 1D pressure model was later improved and the calculated results presented a greater agreement with the experimental data. Song et al. [8] investigated the free surface vortex by establishing a theoretical model of the pressure fluctuation induced by the linear vortex according to the Biot–Savart Law. There was a great agreement between the results of theoretical analysis and model experiments.

Although the method of 1D models can be used to increase the calculation speed as a good compromise of solution accuracy and computation feasibility, experimentation and computational fluid dynamics (CFD) are the most important ways for researchers to investigate hydraulic problems. With the development and wide use of CFD, scholars normally validate the two against each other to increase the credibility of the study and capture the complex flow features in hydraulic machines [9]. Shi et al. [10] designed an axial-flow pump runner by using the method of the surface element based on plane cascade theory. The hydraulic performances of the impeller, calculated by the CFD and model test, showed great agreement with each other. Shi et al. [11] optimized the comprehensive performance of the axial-flow pump with the objective of light weight and high efficiency. The reliability was verified by CFD and model tests, which proved that the accuracy of the approximate model used was high. Yang et al. [12] studied the impact on the hydraulic performance caused by different tip clearance sizes of a sewage pump. Meanwhile, the investigation of the complex vortices structure and propagation benefited from the 3D numerical simulation and test verification. The pressure fluctuation is an important characteristic that reflects the operating state of the water pump. The method of CFD is used to study the pressure fluctuation, which is an important way for the study of the stable operation of the pump. Liu et al. [13] investigated the flow and external characteristics of the pump mode turbine in the hump region, based on the *SST* turbulence model, and proved that the rotating speed determines the characteristics of pressure fluctuation at the wave trough of the hump region. Andreas et al. [14] implemented comparative research on the highly transient flow field in single and double-vane pumps. Based on the mutual verification of experiments and numerical simulations, the characteristics of flow rate and pressure fluctuation on single and double-vane pumps were summarized. Song et al. [15,16] studied the formation mechanism and the dynamic characteristics of the free surface vortex in the pump sump and the effect on the performance of the pump by numerical simulation.

In recent years, many domestic and foreign scholars have carried out a large number of studies on the variable-speed transient process of fluid machinery, but they mainly focused on the unit start-up and shutdown process. Tsukamoto et al. [17] developed theories to predict the transient characteristics of the centrifugal pump during deceleration and proved that the difference between dynamic and quasi-steady characteristics mainly come from impulsive pressure and lag around the vanes. Liu et al. [18] numerically simulated the rapid stopping process of the pump. They found that at the beginning of deceleration, the pressure fluctuation of the pump decreased but the pressure at the inlet increased. Meanwhile, the deviation in the quasi-steady and transient calculation results was attributed to the difference in the internal vortex of the impeller. Chalghoum et al. [19] investigated the dynamic characteristics of the centrifugal pump start-up process under different valve openings and the influence of impeller parameters on pressure changes. Zhang et al. [20,21] numerically investigated the system performance and affinity issue of the tubular pump under variable rotating speed operations. He proved that the similarity law cannot predict the relationship between head and discharge rates during the speed change process completely accurately, but within a certain speed range, the predicted error is acceptable.

However, some characteristics of the large-scale bulb tubular pump unit under the variable-speed operation, such as the pressure fluctuation and distribution, have not been analyzed. There is still room for studies. Therefore, this paper aims to investigate those characteristics of the tubular pump unit, under the operation of rotating speed deceleration (from 1223 rpm to 978.5 rpm), by the methods of CFD and model test. Section 2 describes the computational model and numerical method. Section 3 compares the data of the model test and CFD. The accuracy and reliability of numerical results are verified. The variable speed characteristics calculated by numerical simulation are presented and analyzed in Section 4, and Section 5 summarizes some important conclusions.

2. CFD Method

2.1. Computation Module

The prototype pump in the South-to-North Water Diversion Project with the S-shaped blades is 3350ZGQ37.5-2.45 and the corresponding rated power is 2200 kW. Taking into consideration the exited model test data, this paper established the horizontal bulb tubular pump model components, including inlet conduit, impeller, guide vanes, and outlet conduit, by UG12.0. The main parameters of the bulb tubular pump model are listed in Table 1. The computational fluid region was the entire unit from the inlet section of the inlet conduit to the outlet of the outlet conduit, and the 3D model of the horizontal bulb tubular pump is shown in Figure 1.

Table 1. Characteristic parameters of pump model flow system.

Parameters	Value
Diameter of impeller/mm	315
Number of impeller blades/-	3
Number of guide vanes/-	5
Number of front support vanes/-	6
Blade angle/°	0
Design head/m	2.45
Design discharge/m^3/min	19.9
Initial rotating speed/r/min	1223
Target rotating speed/r/min	978.5 (20% deceleration)

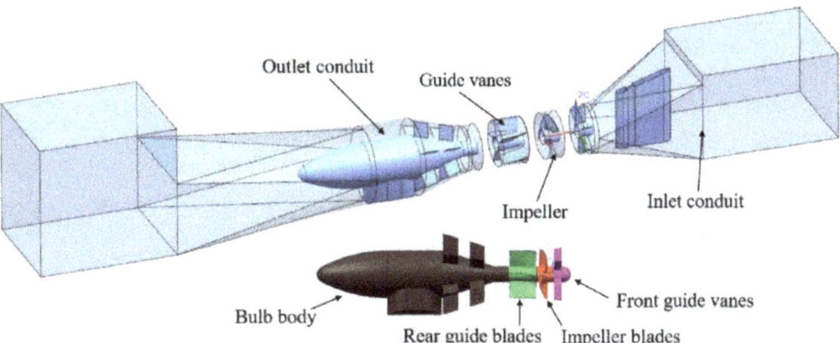

Figure 1. 3D model of horizontal bulb tubular pump.

2.2. Hexahedral and Tetrahedral Mesh

Each part of the computational fluid domain was spat first and then assembled in ANSYS CFX after meshing. In order to take into consideration the later calculation efficiency and accuracy, the hexahedral mesh shown in Figure 2 was generated by using ANSYS ICEM CFD for the impeller portion of the computational model. By controlling the nodes on each topology line, the grid encryption degree of the impeller part is ensured, which is shown in the darker part of Figure 3. The other tetrahedral mesh of the calculation domain was generated by ANSYS MESH. In this paper, the grid independence test was carried out on the grid number of the inlet conduit of the large-scale bulb tubular pumping station during the stable operation under the design conditions. The mesh of the inlet conduit was generated with the same method to ensure that the number of grids changed without reducing quality. Through the steady calculation, it was found that when the number of grids in the calculation domain reached about 4.3 million (Figure 3), the change in hydraulic loss was controlled within $\pm 5\%$, which met the requirement of the grid independence test. As shown in Figure 4, the total number of grid cells in the fluid region was 12,261,547, of which the number of grid cells in the impeller part was 1,760,000.

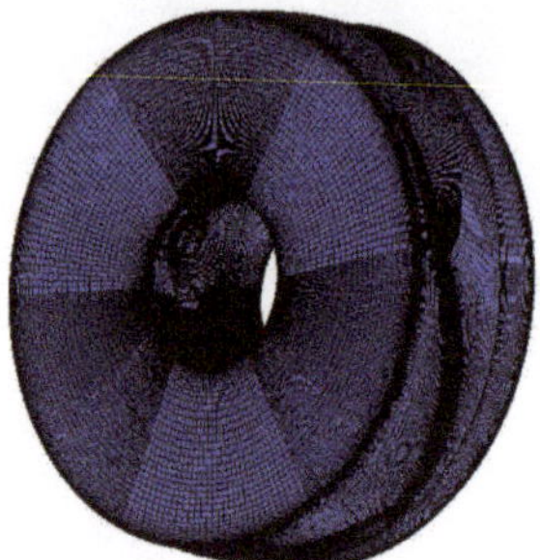

Figure 2. Grid of impeller domain.

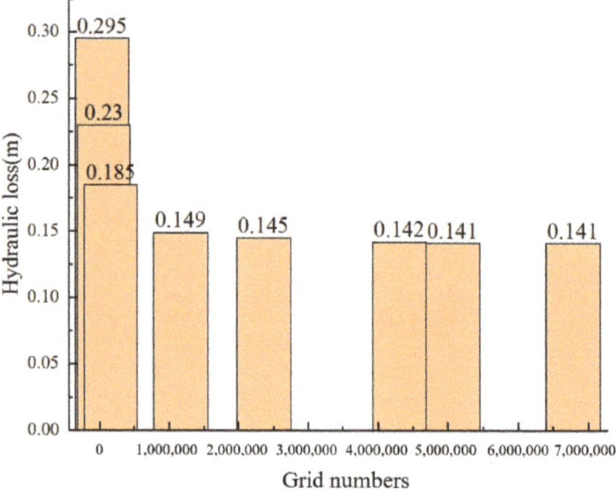

Figure 3. Hydraulic losses under different grid numbers.

Figure 4. Grid of full channel calculation domain.

2.3. Computational Setup

Both a steady and transient 3D simulation of this large-scale bulk tubular pump was conducted by ANSYS CFX based on Reynolds-averaged Navier-Stokes equations (RANS). The results of the steady simulation were calculated first and taken as the initial value of the unsteady simulation. It is important to select an appropriate turbulence model to simulate the complex flow field inside the pump unit. Taking into consideration the compromise of solution accuracy and computation feasibility, the standard *k*-ε model can capture the internal transient flow characteristics of the pump under speed change operation well [22]. Therefore, this one was selected as the turbulence model to close the governing equations, and the SIMPLEC algorithm was used to solve the discrete algebraic equations.

The "Frozen Rotor" interface was used to connect the rotating domain and the stationary domain in the steady numerical simulation, and the "Transient Frozen Rotor" mode was used to transmit data through the interface based on interpolation in the transient numerical simulation. All solid walls were specified with smooth and nonslip, and the calculation medium was normal-temperature water. The boundary condition at the pump outlet was set as total pressure, and the mass flow rate was specified at the pump inlet.

What the paper focused on was the variable-speed characteristics, and it should be carefully considered when the rotating speed of impeller was set. It can be assumed that the efficiency of the pump unit does not change when the pump operates in a certain rotating speed range [19]. Based on the above assumption of constant efficiency, the pump performance parameters at variable speeds are shown as follows:

$$\begin{cases} \frac{Q_1}{Q_2} = \frac{n_1}{n_2} \\ \frac{H_1}{H_2} = \left(\frac{n_1}{n_2}\right)^2 \\ \frac{P_1}{P_2} = \left(\frac{n_1}{n_2}\right)^3 \end{cases} \tag{1}$$

where Q_i, P_i, H_i, and n_i are the flow rate, shaft power, efficiency, and speed of the pump under the working condition of *i*, respectively. It is obvious that the flow rate at the pump inlet will reduce in accompaniment with the rotating speed deceleration. According to the above equations, under the assumption of constant efficiency, the discharge rate of the unit could be achieved by

$$Q = Q_d \times \frac{n}{n_d} \tag{2}$$

where Q_d and n_d are the discharge rate and rotating speed of pump under the design condition, respectively, n is the pump rotating speed under running, and Q is the discharge rate corresponding to n.

Based on the above theories, the algorithm of variable speed was implemented at the rotating domain, through the CFD secondary development (CFX Expression Language): the impeller rotating speed change was set as uniform deceleration from 1223 rpm to 978.5 rpm and the costed time was 1.5 s. The relationship between the rotating speed and the mass flow rate set at the inlet is shown in Figure 5. According to the curve characteristics of Figure 6 in Reference [23], in the first half of the deceleration, the rotating speed and flow rate showed the same trend, conforming to Equation (1), so the setting of the inlet boundary conditions was feasible. In the transient simulation, the total time was set to 3.5424 s, and each time step was set to 0.00246 s, which corresponds to an impeller rotation of 18 deg. The maximum iteration number in each time step was set to 20, and the residual convergence accuracy was set to 10^{-4}.

2.4. Methods of Pressure Fluctuation Analysis

As the pressure fluctuation during the speed change is an unsteady signal, the Fast Fourier Transform (FFT) with a fixed-size analysis window cannot fully reflect the time and frequency characteristics. Therefore, the wavelet method was used to analyze the unsteady pressure fluctuation signal in the process of speed change of the model pump in this paper.

Compared with the method of FFT, wavelet analysis has a good ability to characterize signals in both time and frequency domains, and has a better adaptability to signals [24].

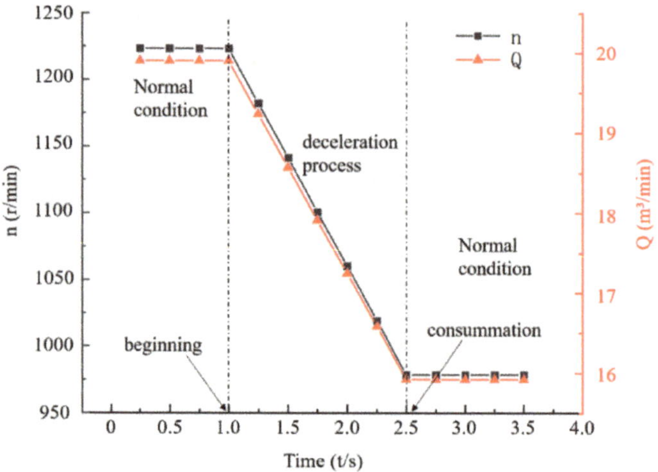

Figure 5. Impeller speed and flow inlet change.

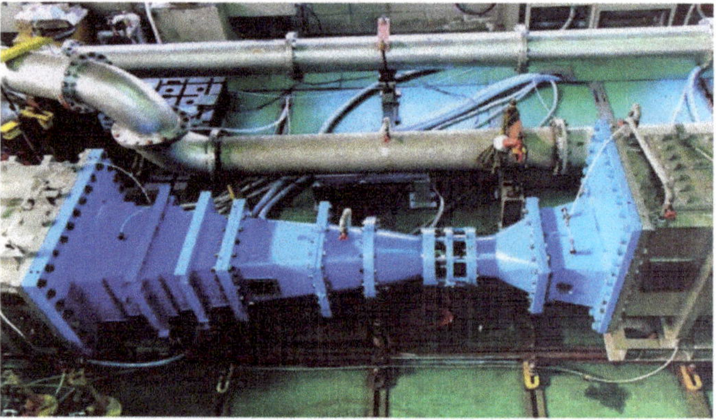

Figure 6. Experimental setup.

3. Comparison of CFD with Experimental Results

3.1. Experimental Device

In order to verify the accuracy of the numerical simulation, the transient external characteristics of the pumping station model are measured by building the bulb tubular pump experimental device system, as shown in Figure 6. In the test, the rotating speed of the model pump is measured by an electromagnetic tachometer. The measurement error of the instrument is less than 0.1%. The discharge of the model device is measured by an electrical flowmeter, and the error is controlled below ±0.15%, the same as the error of the torque measuring device. Hereinafter, the relevant data are processed by a Computer-Aided Measurement System (CAM). According to the complexity of adjusting the speed of the model pump and collecting relevant calculation data during operation, the accuracy of numerical simulation, by comparing the correlation between the steady numerical simulation of the model pump and the external characteristic curve of the experimental simulation under the rotating speed of 1223 r/min, can be verified. The subsequent analysis is carried out through the numerical simulation data.

3.2. Comparison of External Characteristic Curves

Some of the external characteristic curves of the pump are obtained through the model test, and they are compared with the results of the numerical simulation to verify the accuracy and reliability of the subsequent numerical simulation of the pump during the deceleration process. As shown in Figure 7a, the head and efficiency under different flow rates calculated by steady numerical simulation with no cavitation are highly consistent with those achieved by the model test, and the error is small. Considering that the subsequent calculations for deceleration are transient simulations, Figure 7b compares the efficiencies obtained from steady and transient numerical simulation. It can be seen that the two efficiency curves maintain a high degree of consistency. Overall, the investigation of the characteristics of the tubular pump during the deceleration process is reliable using the method of transient numerical simulation.

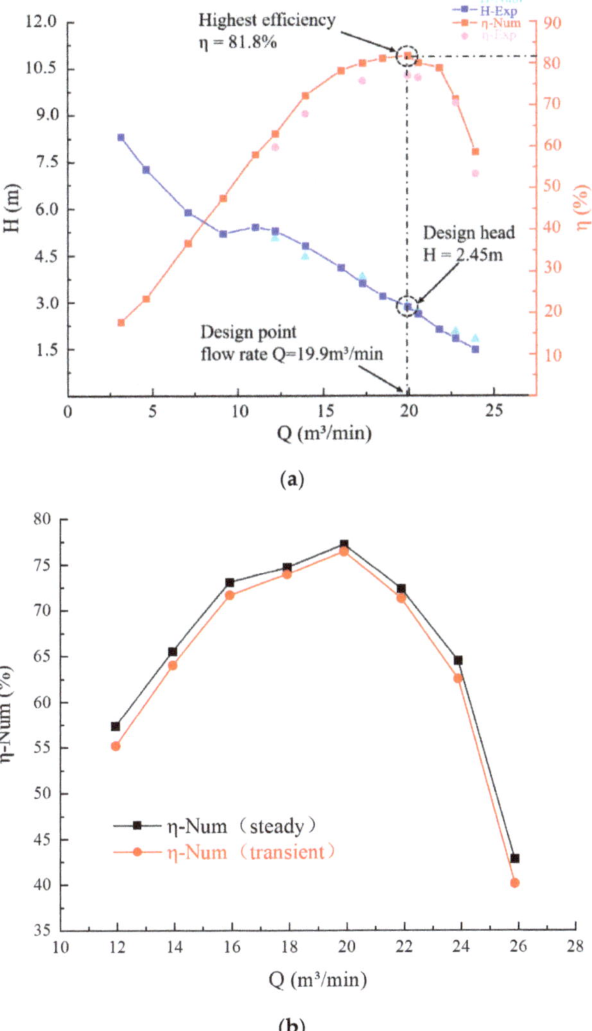

Figure 7. Comparison of numerical simulation and experimental results: (**a**) Steady numerical simulation and experimental results; (**b**) Steady and transient numerical simulation results.

4. Results and Discussion

4.1. Transient External Characteristics of CFD

The transient external characteristic curve of the numerical simulation of the model pump unit during deceleration is shown in Figure 8. The three curves changing with time are pump rotating speed, unit head (the difference between inlet and outlet section pressures), and unit efficiency (the full flow field). It can be seen from the figure that the change in head curve maintains the same trend with the rotating speed. When the model pump runs at 0–1 s, the rotating speed in the rotation domain remains stable. At this period, the head of the model pump is stable at about 2.8 m, and the efficiency is stable at 76%. However, at the beginning of the speed change (1 s), the head curve has an obvious impact phenomenon, and its value is about 2% of the head when the rotating speed is constant at 1223 r/min. This is mainly because, under the transient numerical simulation, the fluid is affected by the inertial force. It can be considered that the pump performance curve does not change at this time; however, the inlet flow begins to decrease, so there is a certain impact head phenomenon; in the 1–2.5 s operation of the model pump, the head of the model pump decreases linearly with the decrease in rotating speed, while the efficiency of the model pump remains stable after increasing to 79% in 0.25 s, an improvement of about 3 percentage points. Starting from the 2.5 s of the model pump operation, the rotating speed of the rotation domain is stable at 987.5 r/min. At this stage, the head and efficiency calculated by CFD are affected by the fluid inertia, resulting in an obvious hysteresis effect, and stabilizes at about 1.77 m and 75% when the time reaches 2.75 s.

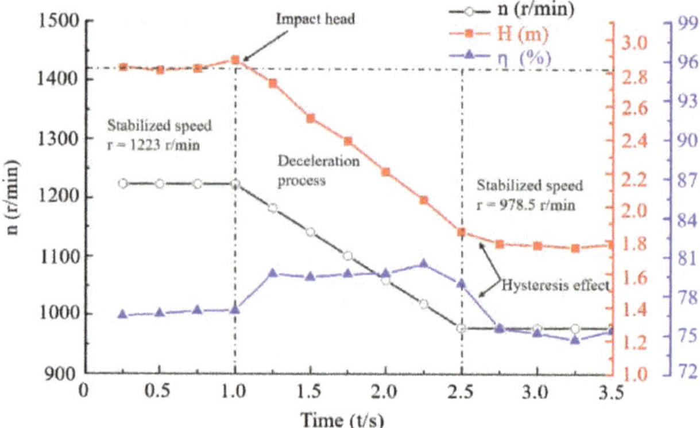

Figure 8. Transient external characteristic curve.

4.2. Analysis of Internal Flow Characteristics

4.2.1. Internal Flow Characteristics in Horizontal Plane

In order to analyze the transient flow field of the tubular pump model in the process of speed change, as shown in Figure 9, the streamline distribution and pressure contours in the horizontal plane of the computational domain at the times of 0.5 s, 1. 5 s, 2.0 s, and 3.0 s are plotted. At 0.5 s, when the rotating speed is stable at 1223 r/min (Figure 9a), the streamlines in the horizontal plane of the entire computational domain are relatively smooth. The pressure distribution of the inlet conduit and the outlet conduit section is relatively uniform, only a comparatively low pressure area appears in the front part of the bulb body, and its value is about 950 kPa. After the water flow is pressurized by the impeller, the pressure distribution of the outlet conduit is obviously larger than that of the inlet conduit. In the process of the rotating speed deceleration (Figure 9b,c), the streamline in the horizontal plane of the water outlet conduit begins to become disordered, and there are turbulence and backflow vortex phenomena at the rear end of the bulb body. The pressure in the inlet conduit begins to increase with time, but the pressure distribution is still uniform, and the comparatively low pressure area in the outlet conduit spreads to about 0.2 m behind the bulb body. After the deceleration is completed (Figure 9d), the streamline in the horizontal plane of the outlet conduit tends to be stable again, but there is still a small amount of vortex and backflow at the rear end of the bulb body, which causes the efficiency reduction shown in Figure 8. The pressure distribution of the outlet conduit is more uniform than that during the rotating speed deceleration, and the comparatively small pressure distribution area is also concentrated only in the front part of the bulb body.

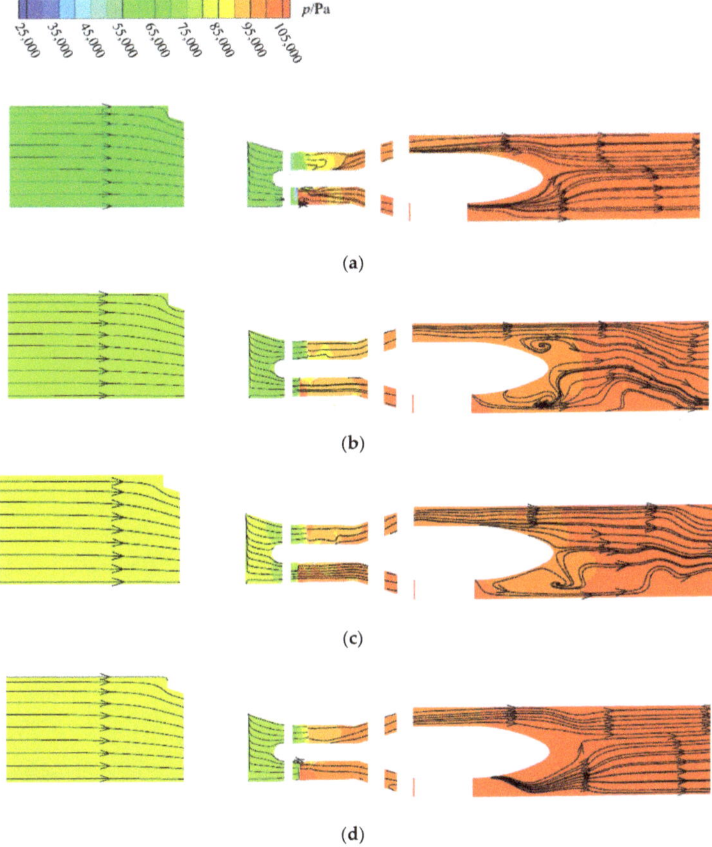

Figure 9. Streamline and pressure contours in horizontal plane of the pump at different times: (**a**) t = 0.5 s; (**b**) t = 1.5 s; (**c**) t = 2.0 s; (**d**) t = 3.0 s.

4.2.2. Pressure Distribution of Impeller

Figure 10 shows the transient pressure distribution contours of the impeller blades at different times. At the time of 0.5 s when the speed is stable at 1223 r/min (Figure 10a), it can be clearly observed that comparatively high pressure is generated on the water inlet edge of the impeller blade pressure side due to the impact of the flow, and the comparatively low pressure is distributed from the hub to the outlet edge of the impeller blades. The pressure side of the entire blade basically shows a trend of slow decline in pressure from the water inlet edge to the outlet edge, but the pressure area distribution is irregular. At the same time, the suction side of the blades shows an obvious pressure stratification: on the water inlet edge of the suction surface, a local negative pressure is generated due to the off-flow. Meanwhile, the pressure distribution on the suction side changes uniformly. The pressure areas of 25 kPa, 35 kPa, and 45 kPa are evenly distributed in the middle of the blade, and the area is about 25% of the entire suction surface. The flow forms a comparatively high pressure area on the water outlet edge of the blade again. In the transition process of the speed change (Figure 10b,c), the pressure value and distribution on the pressure surface of the impeller blade do not change obviously; only the comparatively low pressure area on the water outlet edge and the comparatively high pressure area on the water inlet edge are reduced. On the suction side of the blade, although the pressure value still shows a trend of increasing along the flow direction and the stratification phenomenon is obvious, the pressure of about 50 kPa and the comparatively high pressure area increase significantly. Those pressure areas are about 40% to 60% of the entire suction surface. At 3.0 s after the completion of the speed change (Figure 10d), the area of the comparatively high pressure on the water inlet edge and the comparatively low pressure on the water outlet edge of the suction side are reduced to a minimum. In addition, the pressure of about 50 kPa on the suction surface spreads to about 70% of the entire blade, and the comparatively high pressure area also increases to a maximum.

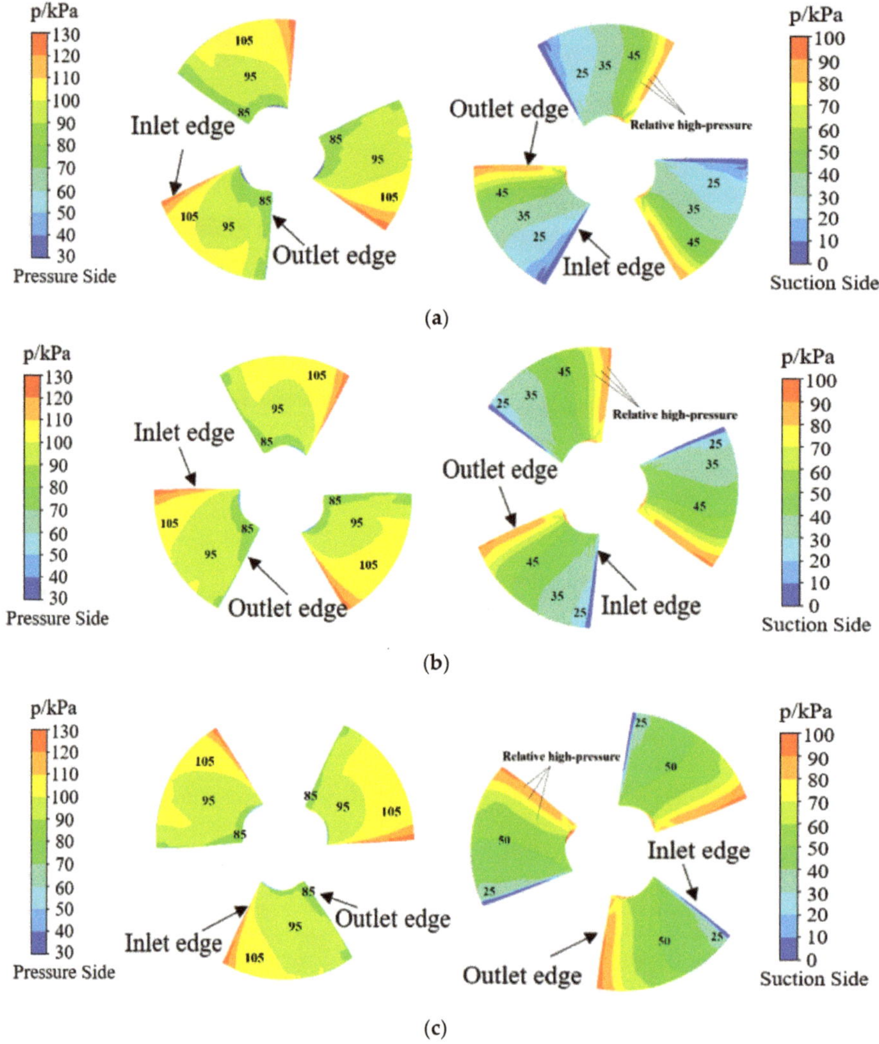

Figure 10. *Cont.*

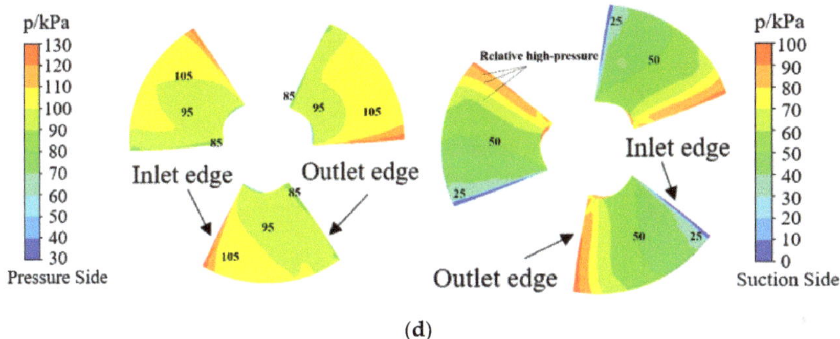

(d)

Figure 10. Pressure distribution contours of the pump impeller blades at different times: (**a**) t = 0.5 s; (**b**) t = 1.5 s; (**c**) t = 2.0 s; (**d**) t = 3.0 s.

In order to show the characteristics of pressure variation along the flow direction on the blade more clearly, three span lines are made along the direction of the hub to the rim, as shown in Figure 11. The pressure distributions at different times on different span lines are drawn in Figure 12. The X-axis represents the comparative locations on the blade, and the Y-axis represents the pressure values corresponding to the comparative locations. It can be obviously observed that the pressure on the blade surfaces present the trend of uniform increase during the process of deceleration. On the two span lines near the middle of the blades and the rim (span 0.5 and span 0.9), the pressure change mainly occurs on the suction side and the inlet edge of the pressure side of the blade, while the pressure near the hub (span 0.5) shows an obvious numerical change on the entire blade surface except the inlet edge of the pressure side.

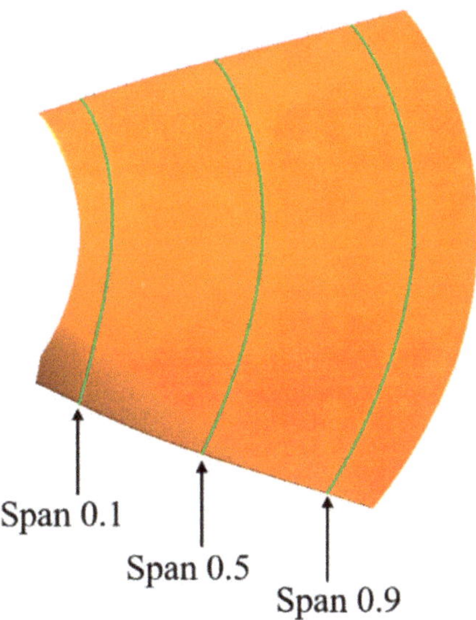

Figure 11. Schematic diagram of the different span line positions.

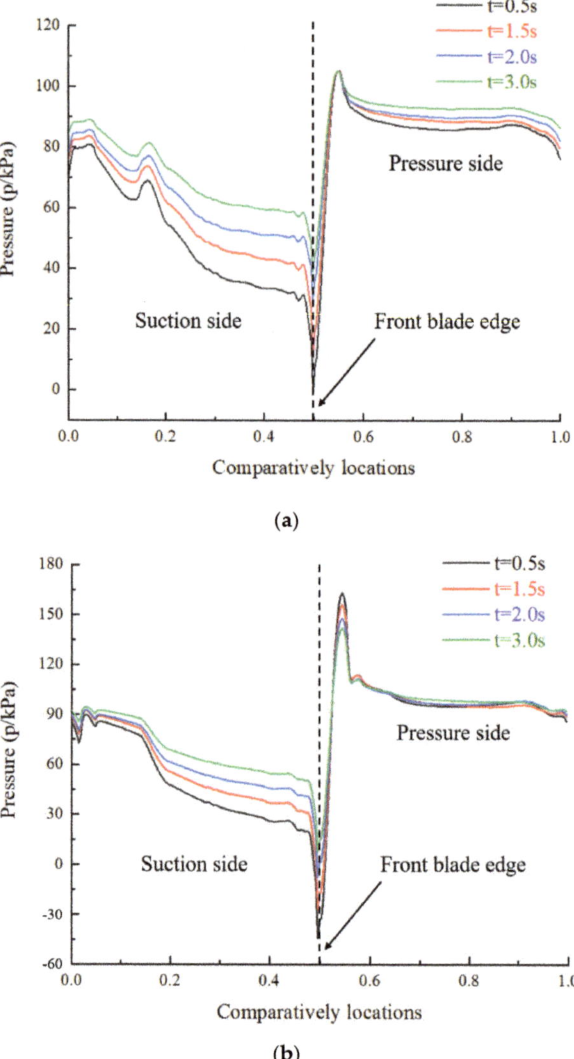

Figure 12. *Cont.*

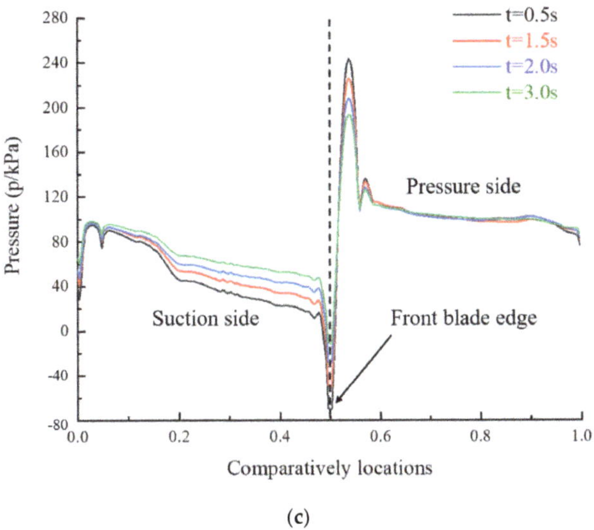

(c)

Figure 12. Pressure characteristics at different span lines under deceleration: (**a**) span = 0.1; (**b**) span = 0.5; (**c**) span = 0.9.

4.3. Pressure Fluctuation

4.3.1. Time Domain Analysis

In order to monitor the pressure fluctuation in the flow field, two monitoring surfaces are set at the front of impeller and the rear of guide vane. In order to capture the influence of radius, three monitoring points are set at the different radii (near the hub, medium radium, and near the rim) of the same monitoring surfaces as those shown in Figure 13.

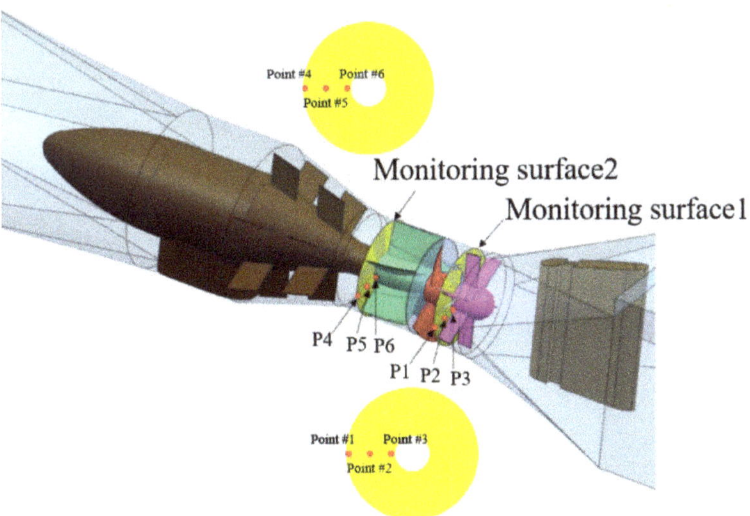

Figure 13. Pressure fluctuation monitoring points.

Figure 14 illustrates the pressure fluctuation time-domain diagram of different monitoring points shown in Figure 13. It can be seen from the figure that the pressure of each monitoring point is quite different from each other: the pressure of the monitoring point at the rim on the impeller inlet section (P1) is smallest, and it increases from 47 kPa to 68 kPa with the deceleration of rotating speed. The closer to the hub, the larger the pressure, but this change decreases as the radius decreases. The pressure of the P3 monitoring point near the hub increases from 157 kPa to 176 kPa with the deceleration of rotating speed. The outlet section of the guide vane also follows the above laws, but the variation is smaller. The data of the two pressure fluctuation monitoring points near the hub are very close, rising from 194 kPa to 197.5 kPa. From the waveform point of view, the pressure fluctuation of the impeller inlet section and guide vane outlet section presents a regular sine or cosine wave, and there are five wave peaks and five wave troughs in a cycle, which is related to the number of blades and guide vanes in the calculation model.

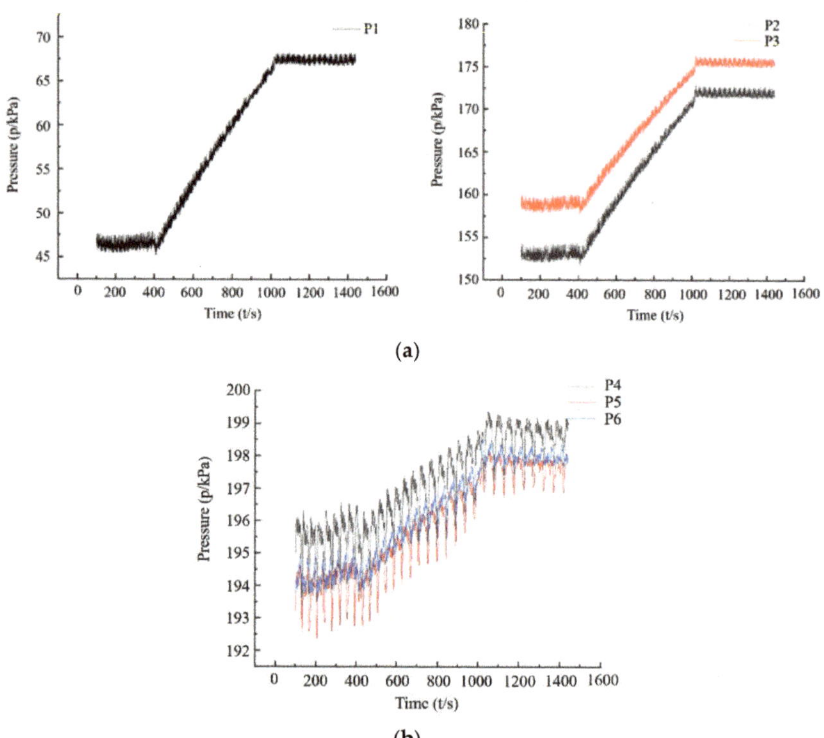

Figure 14. Pressure characteristics of sections: (**a**) impeller inlet section; (**b**) guide vane outlet section.

From the overall trend, in the decreasing process, the pressure fluctuation changes of the monitoring points on the two sections are consistent: with the speed decreasing, the pressure fluctuation value increases linearly, the period becomes larger, and the pressure fluctuation amplitude decreases. However, the pressure change range at the outlet section of the guide vane is only about 18% of that of the corresponding monitoring point at the inlet section of the impeller. When the rotating speed starts to change (1 s), a certain degree of pressure impact appears on two sections [25], while the outlet section of the guide vane also has a certain degree of pressure impact when the change in rotating speed ends (2.5 s). However, compared with the pressure fluctuation when the rotating speed is constant, the value of the impact is comparatively small and can be ignored.

4.3.2. Wavelet Frequency Domain Analysis

As the numerical simulation adopts the linear deceleration method and the number of the impeller blades is 3, the expressions of shaft frequency f_z and blade frequency f are

$$\begin{cases} f_z = \frac{n}{60} \\ f = 3f_z \end{cases} \tag{3}$$

where n is the speed of the pump operation stage, r/min. It can be seen from formula (3) that under the design flow condition, the shaft frequency of the bulb tubular model pump is about 20.38 Hz when the rotating speed is 1223 r/min, and the blade frequency is about 61.14 Hz. When the rotating speed is 987.5 r/min, the shaft frequency is about 16.46 Hz and the blade frequency is about 49.38 Hz. The wavelet time–frequency domain conversion is realized by the "cwt" function of MATLAB, and the time–frequency distribution characteristic map of the pressure fluctuation of different sections is drawn as shown in Figure 15, where the abscissa represents the number of time steps, the ordinate represents the fluctuation frequency, and the color represents the fluctuation amplitude. It can be clearly seen from the figure that the main frequency of pressure fluctuation at the inlet section of the impeller decreases linearly from about 122 Hz to about 100 Hz during the deceleration. It can be found that the calculated values of fluctuation frequency are twice the theoretical values of blade frequency calculated by Equation (3), which is due to the influence of the impeller front support vanes [26]. Meanwhile, the fluctuation amplitude before and after the deceleration is also reduced linearly by about 50%, which is basically in line with the analysis above; the pressure fluctuation energy at the outlet section of the guide vane is mainly concentrated in the low-frequency region of 20 Hz and 10 Hz, and the frequency and amplitude change of the fluctuation before and after the deceleration are not obvious, which is mainly affected by the dynamic and static interference of the impeller and guide vanes. It can be seen from the above analysis that the frequency and amplitude of pressure fluctuation are proportional to the rotating speed, and the frequency characteristics of pressure fluctuation on different sections have obvious differences: the fluctuation in the impeller inlet section is mainly concentrated in the high-frequency region, which is sensitive to the change in rotating speed. The pressure fluctuation of the guide vane outlet section is mainly concentrated in the low-frequency region, and it is not sensitive to the change in rotating speed.

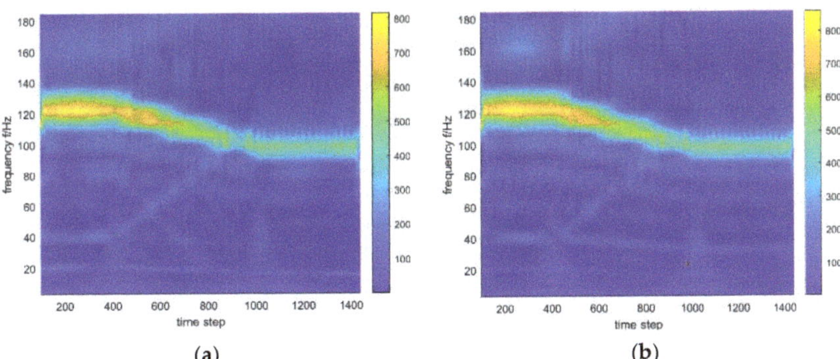

Figure 15. *Cont.*

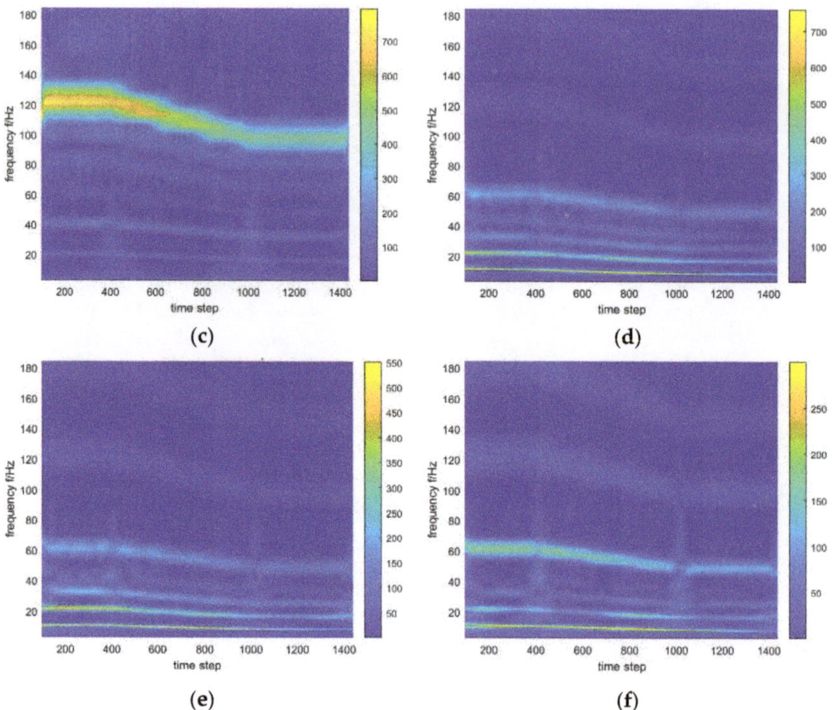

Figure 15. Frequency distribution of pressure fluctuation under two sections: (**a**) monitoring point P1 on the impeller inlet section; (**b**) monitoring point P2 on the impeller inlet section; (**c**) monitoring point P3 on the impeller inlet section; (**d**) monitoring point P4 on the guide vane outlet section; (**e**) monitoring point P5 on the guide vane outlet section; (**f**) monitoring point P6 on the guide vane outlet section.

5. Conclusions

In this paper, the characteristics of the internal and external flow field and the pressure fluctuation of the bulb tubular pump unit during the deceleration are extracted by numerical simulations. The time–frequency domain analysis method based on wavelets is used to investigate the pressure fluctuation obtained. The results provide a certain reference for the energy utilization and safe operation of the water pump unit in adjusting speeds with variable frequency.

(1) The predicted head and efficiency of the pump unit based on the numerical simulation are basically consistent with the experimental results, indicating the reliability of the CFD method. The predicted head curve of the bulb tubular pump based on the unsteady flow field calculation maintains a linear downward trend in the process of deceleration, and there is an impact head phenomenon when the speed begins to change, which is about 2% of the value under the speed of 1223 r/min. The predicted efficiency curve maintains a relatively stable high efficiency in the process of speed reduction, and the efficiency is increased by about 3% compared with the stable condition before the speed change. The two prediction curves have a hysteresis effect of about 0.25 s at the end of the speed change.

(2) In the process of frequency conversion and deceleration of the tubular pump, the pressure distribution on the suction surface of the impeller blade has obvious differences, while this change on the pressure surface is less prominent. At the same time, in the

transition process of deceleration, the pressure distribution on the impeller blades is a regular transition, and there is no sudden change or other characteristics.

(3) From the time-domain analysis of pressure fluctuation, it can be seen that the pressure on the impeller inlet section is sensitive to the change in radius, and the smaller the radius, the smaller the pressure change. Meanwhile, the pressure on the guide vane outlet section is less responsive to the change in radius. With the decrease in rotating speed, the pressure values on the impeller inlet and guide vane outlet sections show a linear upward trend, but the change range of the guide vane outlet section is only about 18% of that on the impeller inlet section. The pressure fluctuation of the two sections has a pressure impact phenomenon at the beginning of the speed change, but the value is small.

(4) From the frequency domain analysis of pressure fluctuation, it can be seen that the impeller inlet section can better reflect the basic characteristics and changing trend of the fluctuation signal than the guide vane outlet section: the pressure fluctuation energy on the impeller inlet section is mainly concentrated in the high-frequency region. Before and after the deceleration, the main frequencies of the fluctuation are 122 Hz and 100 Hz, which are twice the theoretical rotation frequency of 1223 r/min and 987.5 r/min, respectively, showing an obvious linear decreasing trend in the frequency domain characteristic map. Meanwhile, the amplitude of the pressure fluctuation also increases with the pressure fluctuation energy. The energy on the outlet section of the guide vane is mainly concentrated at about 20 Hz and 10 Hz, the difference between the frequencies is not obvious, due to the dynamic and static interference of the impeller and guide vane, and the change in the speed has less of an effect on the fluctuation amplitude.

Author Contributions: Data curation, L.C.; formal analysis, J.L.; methodology, F.X. and W.P.; writing—original draft, J.L. and J.S.; writing—review and editing, L.C., J.L., J.Z. and Y.G.; supervision, L.C. All authors have read and agreed to the published version of the manuscript.

Funding: This research was funded by the National Natural Science Foundation of China (grant no. 51779214), A Project Funded by the Priority Academic Program Development of Jiangsu Higher Education Institutions (PAPD), the Key Project of Water Conservancy in Jiangsu Province (grant no. 2020030 and 2020027), and the Jiangsu Province South-North Water Transfer Technology Research and Development Project (SSY-JS-2020-F-45).

Institutional Review Board Statement: Not applicable.

Informed Consent Statement: Not applicable.

Data Availability Statement: Not applicable.

Conflicts of Interest: The authors declare no conflict of interest.

References

1. Li, Z. *Numerical Simulation and Experimental Study of the Internal Flow in Axial-Flow Pump*; Jiangsu University: Zhenjiang, China, 2007.
2. Zhu, H.; Zhang, R. Numerical Simulation of Internal Flow and Performance Prediction of Tubular Pump with Adjustable Guide Vanes. *Adv. Mech. Eng.* **2014**, *6*, 171504. [CrossRef]
3. Xu, X.M.; Lu, Y.J.; Peng, Z.G.; Lu, W. Discussion on the characteristics and selection of large and medium-sized horizontal pump units. *Jiangsu Water Resour.* **2016**, *2016*, 20–24.
4. Liu, C. Researches and Developments of Axial Flow Pump System. *Trans. Chin. Soc. Agric. Mach.* **2011**, *46*, 49–59.
5. Li, C.X. *Study on Bulb Tubular Pump's Transient Process with VFD*; Yangzhou University: Yangzhou, China, 2011.
6. Gu, Y.D.; Li, J.X.; Wang, P.; Cheng, L.; Qiu, Y.; Wang, C.; Si, Q. An Improved One-Dimensional Flow Model for Side Chambers of Centrifugal Pumps Considering the Blade Slip Factor. *J. Fluids Eng.* **2022**, *144*, 91207. [CrossRef]
7. Gu, Y.; Pei, J.; Yuan, S.; Zhang, J. A Pressure Model for Open Rotor–Stator Cavities: An Application to an Adjustable-Speed Centrifugal Pump with Experimental Validation. *J. Fluids Eng.* **2020**, *142*, 101301. [CrossRef]
8. Song, X.J.; Liu, C.; Wang, Z.W. Prediction on the pressure pulsation induced by the free surface vortex based on experimental investigation and Biot-Savart Law. *Ocean Eng.* **2022**, *250*, 110934. [CrossRef]

9. Kan, K.; Zheng, Y.; Chen, H.; Zhou, D.; Dai, J.; Binama, M.; Yu, A. Numerical simulation of transient flow in a shaft extension tubular pump unit during runaway process caused by power failure. *Renew. Energy* **2020**, *154*, 1153–1164. [CrossRef]
10. Shi, L.; Zhang, W.; Jiao, H.; Tang, F.; Wang, L.; Sun, D.; Shi, W. Numerical simulation and experimental study on the comparison of the hydraulic characteristics of an axial-flow pump and a full tubular pump. *Renew. Energy* **2020**, *153*, 1455–1464. [CrossRef]
11. Shi, L.; Zhu, J.; Tang, F.; Wang, C. Multi-Disciplinary Optimization Design of Axial-Flow Pump Impellers Based on the Approximation Model. *Energies* **2020**, *13*, 779. [CrossRef]
12. Yang, Y.; Zhou, L.; Bai, L.; Xu, H.; Lv, W.; Shi, W.; Wang, H. Numerical Investigation of Tip Clearance Effects on the Performance and Flow Pattern Within a Sewage Pump. *J. Fluids Eng.* **2022**, *144*, 81202. [CrossRef]
13. Liu, J.; Liu, S.; Wu, Y.; Jiao, L.; Wang, L.; Sun, Y. Numerical investigation of the hump characteristic of a pump–turbine based on an improved cavitation model. *Comput. Fluids* **2012**, *68*, 105–111. [CrossRef]
14. Pesch, A.; Melzer, S.; Schepeler, S.; Kalkkuhl, T.; Skoda, R. Pressure and Flow Rate Fluctuations in Single- and Two-Blade Pumps. *J. Fluids Eng.* **2020**, *143*, 11203. [CrossRef]
15. Song, X.-J.; Yao, R.; Chao, L.; Wang, Z.-W. Study of the formation and dynamic characteristics of the vortex in the pump sump by CFD and experiment. *J. Hydrodyn.* **2021**, *33*, 1202–1215. [CrossRef]
16. Song, X.; Luo, Y.; Wang, Z. Numerical prediction of the influence of free surface vortex air-entrainment on pump unit performance. *Ocean Eng.* **2022**, *256*, 111503. [CrossRef]
17. Tsukamoto, H.; Matsunaga, S.; Yoneda, H.; Hata, S. Transient Characteristics of a Centrifugal Pump During Stopping Period. *J. Fluids Eng.* **1986**, *108*, 392–399. [CrossRef]
18. Liu, J.; Li, Z.; Wang, L.; Jiao, L. Numerical Simulation of the Transient Flow in a Radial Flow Pump during Stopping Period. *J. Fluids Eng.* **2011**, *133*, 111101. [CrossRef]
19. Chalghoum, I.; Elaoud, S.; Akrout, M.; Taieb, E.H. Transient behavior of a centrifugal pump during starting period. *Appl. Acoust.* **2016**, *109*, 82–89. [CrossRef]
20. Zhang, R.T.; Yao, L.B.; Zhu, H.G.; Zhang, L.; Wei, J. Low-head pumping system performances and affinity issues under variable speed operation based on CFD. *J. Drain. Irrig. Mach. Eng.* **2010**, *28*, 5.
21. Cheng, J.L.; Zhang, R.T.; Deng, D.S.; Yi, D.; Xusong, F. Adaptability research of optimal operation mode with variable frequency drives for pumping stations in China's Eastern Route Project of S-to-N Water Diversion. *J. Drain. Irrig. Mach. Eng.* **2010**, *28*, 5.
22. Li, W.; Ji, L.L.; Shi, W.D.; Zhang, Y.; Zhou, L. Numerical Calculation of Internal Flow Field in Mixed-flow Pump with Non-uniform Tip Clearance. *Trans. Chin. Soc. Agric. Mach.* **2016**, *47*, 66–72.
23. Yang, L.; Xu, Z.; Zhen, Y. Study on shutdown transition process of large low head pumping station. *Water Conserv. Constr. Manag.* **2020**, *40*, 73–79.
24. Zhang, F. Application of Wavelet on Status Monitoring and Failure Diagnosis of Large Scale and Intermediate Pump Unit. Master's Thesis, Hohai University, Nanjing, China, 2007.
25. Li, W.; Lu, D.L.; Ma, L.L.; Ji, L.; Wu, P. Experimental study on pressure vibration characteristics of mixed-flow pump during start-up. *Trans. Chin. Soc. Agric. Eng.* **2021**, *37*, 44–50.
26. Shi, W.; Cai, R.M.; Li, S.B.; Sun, T.; Shen, C.; Cheng, L.; Luo, C. Numerical simulation of pressure fluctuation in postpositional bulb tubular pump. *South North Water Transf. Water Sci. Technol.* **2021**, *37*, 44–50.

MDPI

Article

Numerical Analysis and Model Test Verification of Energy and Cavitation Characteristics of Axial Flow Pumps

Chuanliu Xie [1,*], Cheng Zhang [1], Tenglong Fu [1], Tao Zhang [1], Andong Feng [1] and Yan Jin [2]

[1] College of Engineering, Anhui Agricultural University, Hefei 230036, China
[2] College of Hydraulic Science and Engineering, Yangzhou University, Yangzhou 214000, China
* Correspondence: xcltg@ahau.edu.cn

Abstract: In order to study the energy and cavitation performance of a high-ratio axial flow pump, the SST k-ω turbulence model and ZGB cavitation model were used to numerically calculate the energy and cavitation performance of a high-ratio axial flow pump, and a model test analysis was carried out. The study concluded that the errors in the numerical calculation of head, efficiency, and critical cavitation margin are within 0.2 m, about 3% and 5%, respectively, and the numerical calculation results are reliable. For the flow conditions of Q = 411 L/s, 380 L/s, 348 L/s, and 234 L/s, the numerically calculated critical cavitation margins are 7.1 m, 5.7 m, 4.6 m, and 9.5 m, respectively, and the experimental critical cavitation margins are 7.5 m, 4.9 m, 4.6 m, and 9.5 m, respectively, with errors of −0.4 m, 0.8 m, 0.0 m, and 0.0 m, in that order; numerical calculations and test results trend the same, with small errors. Under the same inlet pressure, as the flow rate decreases, the vacuole first appears at the head of the blade pressure surface under the large flow rate condition (Q = 411 L/s), and the vacuole appears at the head of the blade suction surface under the small flow rate condition (Q = 234 L/s). As the inlet pressure decreases ($p_{in} = 11 \times 10^4$–4×10^4 Pa), the vacuole gradually increases under the same flow rate and the cavitation degree increases. The research results of this paper can provide a reference for the study of the energy and cavitation mechanism of the same type of axial flow pump.

Keywords: axial flow pumps; energy; cavitation; numerical calculation; test

Citation: Xie, C.; Zhang, C.; Fu, T.; Zhang, T.; Feng, A.; Jin, Y. Numerical Analysis and Model Test Verification of Energy and Cavitation Characteristics of Axial Flow Pumps. *Water* **2022**, *14*, 2853. https://doi.org/10.3390/w14182853

Academic Editors: Ran Tao, Changliang Ye and Xijie Song

Received: 23 July 2022
Accepted: 8 September 2022
Published: 13 September 2022

Publisher's Note: MDPI stays neutral with regard to jurisdictional claims in published maps and institutional affiliations.

Copyright: © 2022 by the authors. Licensee MDPI, Basel, Switzerland. This article is an open access article distributed under the terms and conditions of the Creative Commons Attribution (CC BY) license (https://creativecommons.org/licenses/by/4.0/).

1. Introduction

When water pumps are used for high-speed rotating parts in the overflow, in the operation of unreasonable conditions, the overflow part of the local area pressure reduces the vaporization pressure. The liquid then begins to vaporize, generating a large number of bubbles. These bubbles, filled with gas or steam, quickly expand and move with the liquid flow to a higher pressure. The presence of bubbles around the higher liquid pressure causes the bubbles to shrink sharply and quickly condense and collapse, while the bubbles around the liquid mass, due to inertia, fill the bubble at high speed, creating a mutual impact, noise, and vibration, seriously affecting the performance of the pump. Therefore, it is necessary to carry out research on the distribution of cavitation and the pump performance under different degrees of cavitation.

Several scholars have conducted research on cavitation in pumps, and some of them have studied the numerical computational model of cavitation, resulting in the Rayleigh–Plesset equation [1,2], the Kubota equation [3], and the Singhal equation [4] which are better applied in the numerical computation of cavitation. Some scholars have modified the numerical calculation models of cavitation by modifying the turbulent viscosity [5], developing a two-phase three-component cavitation numerical calculation model [6], applying density correction to the turbulence model [7], and modifying the Zwart cavitation model [8], all of which have improved the accuracy of the numerical calculation to some extent. Additionally, some scholars have analyzed the applicability and accuracy of several cavitation models and have concluded that the accuracy of the prediction using the SST (k-ω) turbulence model with the ZGB cavitation model is higher [9–12]. The research of related scholars can provide a reference for the numerical calculation of the cavitation characteristics in this paper.

Related scholars have studied the flow field under pump cavitation using the cavitation model developed by the above-mentioned scholars [13] and have concluded that the cavitation region has a great influence on the velocity field, leading to the degradation of pump performance [14] and revealing that the leakage vortex at the tip of the blade top induces cavitation [15]. Furthermore, some scholars have studied the cavitation characteristics of pumps based on time and frequency domain vibration analysis techniques [16] and acoustic analysis techniques [17] and have investigated the vibration characteristics and pressure pulsation characteristics in pumps under cavitation [18–21]. Yet more scholars have carried out structural optimization and analysis in order to improve the cavitation performance of the pump [22,23]. The studies of related scholars provide references for the experimental tests in this paper.

High efficiency and poor cavitation characteristics are two relatively contradictory quantities. The high-ratio axial flow pump studied in this paper has excellent energy characteristics and cavitation characteristics, and this paper expects to reveal its energy and cavitation characteristics and cavitation flow field distribution. In this paper, firstly, the k-ω SST turbulence model and ZGB cavitation model are used to numerically calculate the cavitation performance of a high specific speed axial flow pump, which is verified by model tests and error analysis. After verifying the accuracy of the numerical calculation, the distribution law of the cavitation under cavitation, which is not easily obtained in the test, is further revealed. The research results can provide a reference for an in-depth exploration of the design and cavitation characteristics of high-specific speed axial flow pumps.

2. Numerical Calculation Models, Grids, and Methods

2.1. Numerical Calculation Model

The axial flow pump design parameters, as shown in Table 1, include the axial pump impeller diameter D = 300 mm, hub ratio d/D = 0.35, four impeller blades, seven guide vane blades, and a blade placement angle of 0°.

Table 1. Axial flow pump design parameters.

Parameters	Numerical Value	Unit
Impeller speed/n	1450	rpm
Design flow/Q_d	350	L/s
Design head/H_d	5.5	m
Specific speed/n_s	872	r/min

The 3D model of the inlet pipe, impeller, guide vane, and outlet pipe is established by Solidworks, and the overall structure is shown in Figure 1.

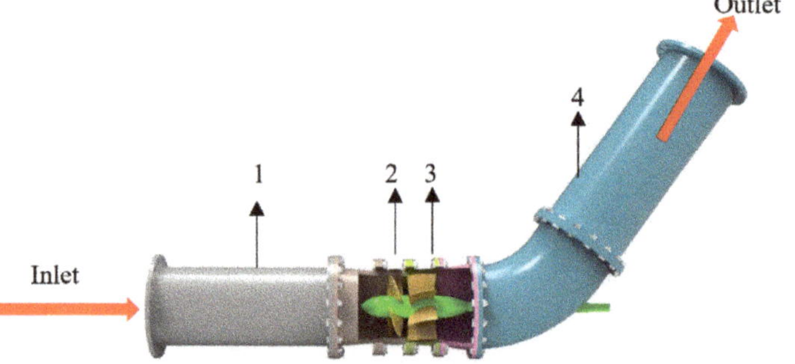

Figure 1. Overall 3D model drawing.1. Inlet pipe, 2. Impeller, 3. Guide vane, and 4. Outlet pipe.

2.2. Mesh Division

The fluid channels were extracted from the 3D model and ICEM was used to structure the meshing of the inlet and outlet pipes. The topology of the inlet and outlet pipes was established separately in the division, and the point-line surface correlation was made one by one. In order to increase the fit of the face of the circular pipe model, we created an "O" type mesh structure, then set the mesh size, considering the thickness of the boundary layer, encrypted the boundary layer locally (the change rate of the mesh from the boundary to the inside is 1.05), and, finally, performed the mesh partitioning, imported the design data files of the impeller and guide vane into TurboGrid for the modeling of the impeller and guide vane, and then used TurboGrid to structure the mesh partitioning of the impeller and guide vane. The overall mesh is shown in Figure 2.

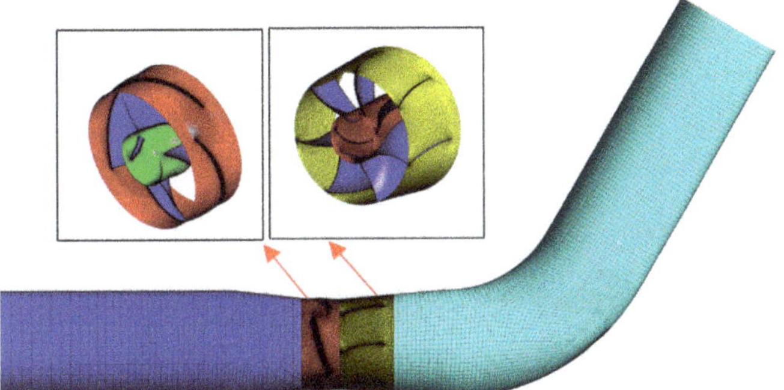

Figure 2. Overall structured grid.

Each component of the computational domain is divided by structured mesh and the mesh quality is above 0.35. The mesh division method and the main parameters are shown in Table 2 (y^+ is a dimensionless quantity of distance from the wall, which is proportional to the height of the first grid layer of the wall. In numerical calculations using SST *k-ω* and RNG *k-ε* turbulence models, the rotational and shear flow y^+ is taken to be 30–100). After the mesh irrelevance analysis, the total mesh number is finally selected as 2114505 for numerical calculation.

Table 2. Grid parameters of main components of axial flow pump.

Parts	Grid Division Method	Average y^+ Value
Impeller	"J" topology	≈50
Guide vane	"O" topology	≈50
Impeller leaf top clearance	"H" topology and 8-layer grid arrangement	<10

2.3. Control Equations and Boundary Conditions

In this paper, the SST *k-ω* turbulence model [24] is used for the numerical energy performance calculation of axial flow pumps, and the SST *k-ω* turbulence model and ZGB (Zwart–Gerber–Belamri) model are used for the numerical calculation of their cavitation performance.

Turbulence control equation (N-S equation):

$$\frac{\partial(\rho u_i)}{\partial t} + \frac{\partial(\rho u_i u_j)}{\partial x_j} = -\frac{\partial P}{\partial x_i} + [\mu(\frac{\partial u_i}{\partial x_j} + \frac{\partial u_j}{\partial x_i})] + F_i \tag{1}$$

where t is time (s); ρ is fluid density (kg/m^3); x_i and x_j are spatial coordinates; u_i and u_j are the velocity components of the fluid parallel to the corresponding axes x_i and x_j, respectively; F_i is the volume force component in the i-direction; μ is the fluid dynamic viscosity coefficient; and p is the pressure (Pa).

The transport equation of the SST k-ω turbulence model can be expressed as:

$$\frac{\partial(\rho k)}{\partial t}+\frac{\partial(\rho k u_i)}{\partial x_i}=\frac{\partial}{\partial x_j}[(\mu+\frac{\mu_t}{\sigma_k})\frac{\partial k}{\partial x_j}]+G_k-Y_k+S_k \tag{2}$$

$$\frac{\partial(\rho\omega)}{\partial t}+\frac{\partial(\rho\omega u_i)}{\partial x_i}=\frac{\partial}{\partial x_j}[(\mu+\frac{\mu_t}{\sigma_\omega})\frac{\partial\omega}{\partial x_j}]+G_\omega-Y_\omega+S_\omega+D_\omega \tag{3}$$

where G_k, G_ω is the generating term of the equation; Y_k, Y_ω is the generating term of the diffusive action; S_k, S_ω is the user-defined source term; D_ω is the term generated by the orthogonal divergence; k is the turbulent kinetic energy; ω is the turbulent special dissipation; and μ_t is turbulent dynamic viscosity coefficient.

Where the interphase mass transfer equation of the ZGB model [25] is

$$\frac{\partial}{\partial t}(f_v\rho)+\nabla\cdot\left(f_v\rho\vec{V_v}\right)=\nabla\cdot(\Gamma\nabla f_v)+R_e-R_c \tag{4}$$

$$R_e=F_{vap}\frac{3\alpha_{nuc}(1-\alpha_v)\rho_v}{\Re_B}\sqrt{\frac{2(P_v-P)}{3\rho_l}}(1-f_v-f_g) \qquad (if\ p\ \le\ p_v) \tag{5}$$

$$R_c=F_{cond}\frac{3\alpha_v\rho_v}{\Re_B}\sqrt{\frac{2(P_v-P)}{3\rho_l}}f_v \qquad (if\ p\ \ge\ p_v) \tag{6}$$

where f_v is the vapor mass fraction, Γ is the diffusion coefficient, Re is the evaporation conversion of the gas-liquid phase, Rc is the condensation conversion of the gas-liquid phase, R_B is the bubble radius, α_{nuc} is the volume fraction of the nucleation site, F_{vap} is the evaporation coefficient, F_{cond} is the condensation coefficient, ρ_v is the vapor density, α_v is the volume fraction of the vapor phase, P_v is the pressure inside the bubble, p is the pressure around the bubble in the liquid, ρ_l is the liquid density, $\vec{V_v}$ is the mode of the relative velocity of liquid and vapor, and f_g is the gas mass fraction.

In the numerical calculation software, the mesh model of each component is imported into CFX-Pre, and the mesh model of each segment is assembled to form the geometric model for the numerical calculation of the axial flow pump; the calculation settings are shown in Table 3.

Table 3. Calculation settings of the main components of the axial flow pump.

Boundary Conditions	Setting Method
Impeller speed	1450 r/min
Inlet	Total pressure
Outlet	Flow
Static wall	Application of no-slip conditions
Near Wall Area	Standard Wall Functions
Dynamic and static interface	The "Stage" interface
Interfaces	GGI grid stitching technology

The inlet condition is set to a total pressure of 11 × 10^4 Pa and the outlet condition is set to a flow rate of Q = 210–434 L/s when performing the energy characteristic calculation, and the inlet condition is set to a total pressure of 11 × 10^4 Pa, 10 × 10^4 Pa, 8 × 10^4 Pa, 7 × 10^4 Pa, 6 × 10^4 Pa, 5 × 10^4 Pa, and 4 × 10^4 Pa and the outlet condition is set to a flow rate of 411 L/s, 380 L/s, 348 L/s, and 234 L/s, respectively, when performing the cavitation calculation. The diffusion term and pressure gradient are represented by finite element functions, the convective term is represented by a high-resolution format (High-Resolution Scheme), and the velocities u, v, and w, in the pressure p, x, y, and z directions of the monitored flow field are calculated; the convergence conditions of turbulent kinetic energy k and dissipation rate ε are set to 10^{-6}, and, in principle, the smaller the residuals, the better.

2.4. Numerical Calculation Results Analysis Formula

The calculation formulas [26–28] for the prediction of the device head, H_{net}, and efficiency, η, of the overflow components based on the flow velocity and pressure fields obtained from numerical calculations are:

$$H_{net} = \left(\frac{\int_{S_2} p_2 u_{t2} dS}{\rho Q g} + H_2 + \frac{\int_{S_2} {u_2}^2 u_{t2} dS}{2Qg}\right) - \left(\frac{\int_{S_1} p_1 u_{t1} dS}{\rho Q g} + H_1 + \frac{\int_{S_1} {u_1}^2 u_{t1} dS}{2Qg}\right) \tag{7}$$

$$\eta = \frac{\rho g Q H_{net}}{T\omega} \times 100\% \tag{8}$$

where H_{net} is the head (m), S_1, S_2 is the area of the inlet and outlet section of the axial flow pump (m^2), P_1, P_2 is the static pressure at each point of the inlet and outlet section of the axial flow pump (Pa), u_{t1}, u_{t2} is the normal component of the flow velocity at each point of the inlet and outlet section of the axial flow pump (m/s), ρ is the density (kg/m^3), Q is the flow rate (m^3/s), g is the acceleration of gravity (m/s^2), H_1, H_2 is the elevation of the inlet and outlet section of the axial flow pump (m), u_1, u_2 is the flow velocity at each point of the inlet and outlet channel section of the axial flow pump (m/s), η is the efficiency (%), T_p is the torque (N·m), and ω is the rotational angular velocity of the impeller (rad/s).

3. Test Device and Test Method

3.1. Test Device

The test bench is a vertical closed circulation system as shown in Figure 3.

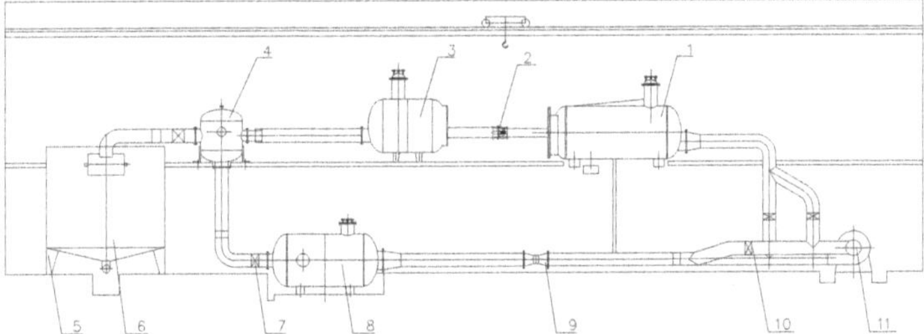

Figure 3. High-precision plan of the hydraulic mechanical test bench. 1. Inlet tank, 2. Test pump, 3. Pressure outlet tank, 4. Bifurcation tank, 5–6. Flow rate in-situ calibration device, 7. Working condition regulating gate valve, 8. Pressure regulating rectifier, 9. Electromagnetic flow meter, 10. Operating control gate valve, and 11. Auxiliary pump unit.

The test head is measured by a differential pressure transmitter (accuracy ±0.015%), the flow rate is measured by an electromagnetic flowmeter (accuracy ±0.18%), the speed and torque are measured by a speed and torque sensor (accuracy ±0.24%), and the cavitation margin is measured by an absolute pressure transmitter (accuracy ±0.015%), with a comprehensive uncertainty of ±0.39% on the test bench.

3.2. Test Methods

The pump head, H, is equal to the total energy head difference between the two pressure measuring sections of the pump inlet and outlet [29].

$$H = \left(\frac{p_2}{\rho g} - \frac{p_1}{\rho g} + z_2 - z_1\right) + \left(\frac{u_2^2}{2g} - \frac{u_1^2}{2g}\right) \tag{9}$$

The shaft power N is calculated by the following Equation [30]:

$$N = \frac{\pi}{30} n(M - M') \quad (10)$$

where H is the pump head (m), P_1, P_2 is the static pressure at the inlet and outlet of the flow field (Pa), z_1, z_2 is the height of the inlet and outlet of the flow field (m), u_1, u_2 is the flow velocity of the inlet and outlet of the flow field (m/s), ρ is the density of the water in real-time of the test (kg/m^3), g is the local acceleration of gravity (m/s^2), N is the shaft power (w), M is the input torque of the pump (N·m), M' is the mechanical loss torque of the pump (N·m), and n is the test speed of the pump (r/min).

During the test, the effective cavitation margin ($NPSH_{av}$) corresponding to a 1% drop in efficiency is defined as the critical cavitation margin ($NPSH_{re}$) when the flow rate is kept constant. The effective cavitation margin value for the pump at different inlet pressures, $NPSH_{av}$, is calculated by the following Equation [31]

$$NPSH_{av} = \frac{p_{av}}{\rho g} + h + \frac{v^2}{2g} - \frac{p_v}{\rho g} \quad (11)$$

where $NPSH_{av}$ is the pump effective cavitation margin (m), p_{av} is the pump into the water tank pressure measurement point of the absolute pressure, measured by the absolute pressure transmitter (Pa), ρ is the test of real-time water density (kg/m^3), g is the local acceleration of gravity (m/s^2), v is the pump into the tank pressure measurement section average flow rate (m/s), P_v is the test water temperature of the water saturation vapor pressure (Pa), and h is the absolute pressure transmitter above the pump vane rotation centerline (pump shaft) height value (m).

The pump efficiency referred to in this paper is the value after deducting the mechanical loss of torque and is calculated by the following formula [32]:

$$\eta = \frac{\rho g Q H}{N} \times 100\% \quad (12)$$

where η is the pump model efficiency (%), H is the pump head (m), Q is the pump flow (m^3/s), ρ is the test real-time water density (kg/m^3), g is the local gravitational acceleration (m/s^2), and N is the shaft power (w).

4. Numerical Calculations and Analysis of Experimental Results

4.1. Numerical Calculations and Experimental Energy Analysis

The numerical calculations and test results for the flow-head and flow-efficiency of high-specific speed axial flow pumps are collated and compared in Figure 4.

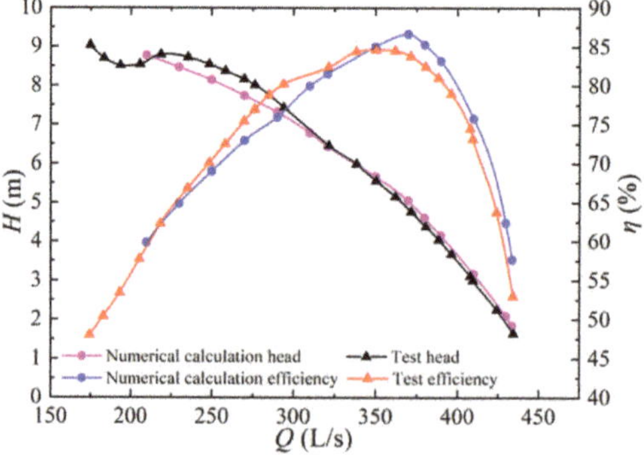

Figure 4. Comparison of the numerical calculations and experimental energy performance results.

Figure 4 shows that the head is H_d = 5.56 m and the efficiency is η = 84.6% when the flow rate is Q_d = 350.39 L/s. The head at the design point meets the design requirements and the efficiency is within the high-efficiency zone, which shows the reasonableness of the design. The highest operating head is H_m = 8.80 m, 1.58 times the design head, indicating that the high specific speed axial flow pump designed in this paper can be operated in a wider range. When the head is greater than 8.80 m, the axial flow pump enters the saddle area (Q = 174.68–218.74 L/s), and the operation becomes unstable, accompanied by an increase in bad flow patterns, vibration, and a sharp increase in noise; operation in this area should be avoided.

Because the test and numerical calculations are not at the same operating point, the interpolation points of the head and efficiency curves (within the range of Q = 210–434 L/s) were found using Origin software. The difference between the tested and numerically calculated heads at the same operating conditions after interpolation was obtained in Figure 5, and the difference between the tested and numerically calculated efficiencies at the same operating conditions was obtained in Figure 6.

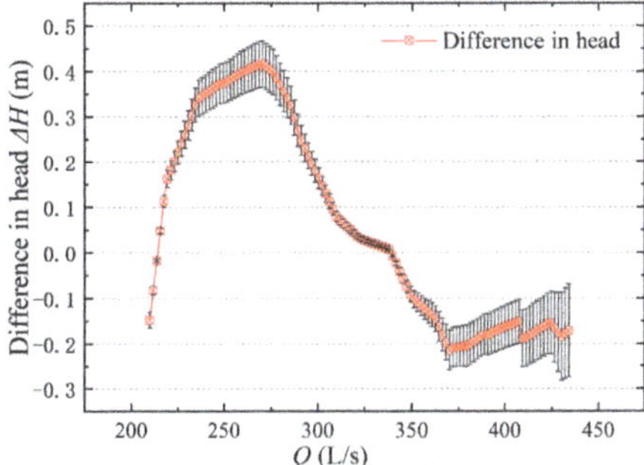

Figure 5. Head error analysis graph.

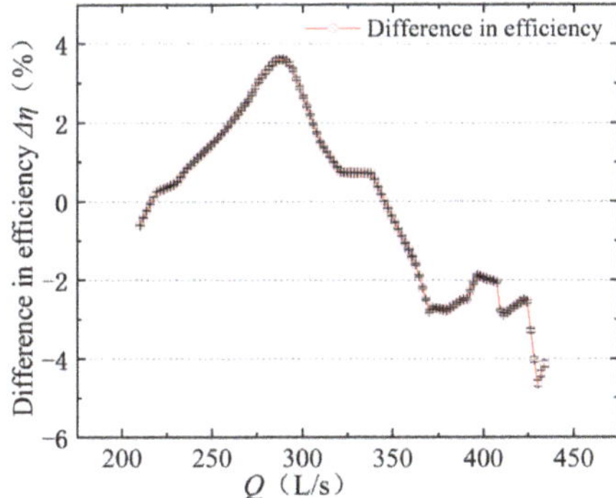

Figure 6. Efficiency analysis graph.

As shown in Figure 5, the error between the numerical calculation and test results for the flow rate-head is small, with the maximum error at the high head being 0.4 m and the basic error range around 0.2 m. The prediction is more accurate in the high-efficiency zone conditions (Q = 293.93–434.53 L/s) and slightly off in the low flow rate conditions (Q = 174.68–293.93 L/s). As shown in Figure 6, the maximum error between the numerical calculations and test results for flow rate efficiency is no more than 5%, with a basic error range of 3% or less. Overall, the numerical calculations have good accuracy in predicting the energy performance of the axial flow pump.

4.2. Numerical Calculations and Experimental Cavitation Properties

The results of numerical calculations and experimental tests of the 0° cavitation characteristics of the high specific speed axial flow pump are collated and compared in Figure 7, where the horizontal coordinate is the inlet pressure and the vertical coordinate is the efficiency curve, with a 1% drop in efficiency for the critical cavitation condition during numerical calculations and tests.

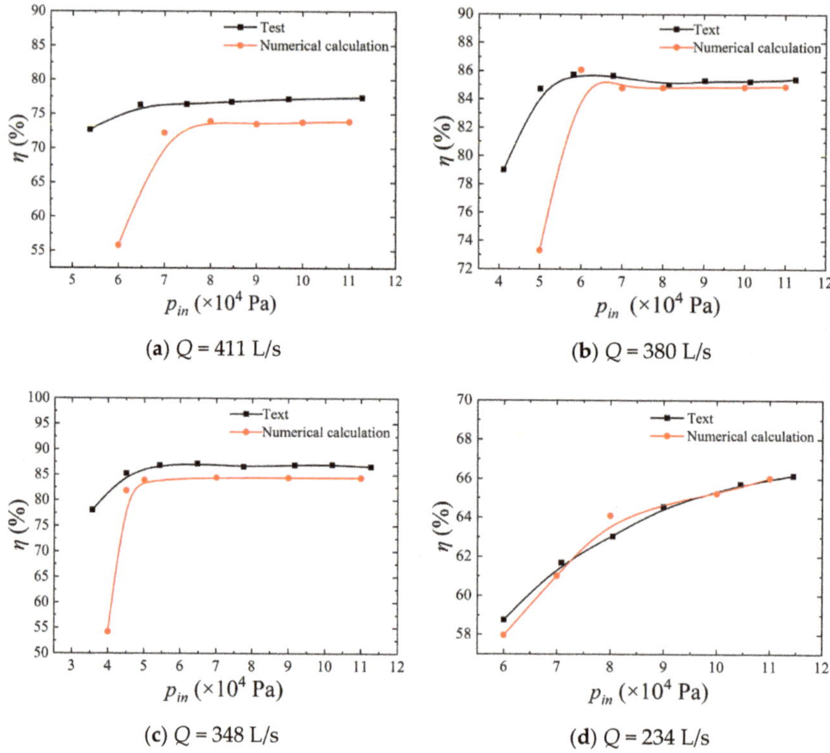

Figure 7. Comparison of the numerical calculation and the test cavitation performance results.

The numerical calculations of the 0° cavitation characteristics were collated with the critical cavitation margins taken out for the different operating conditions of the test to compare in Figure 8.

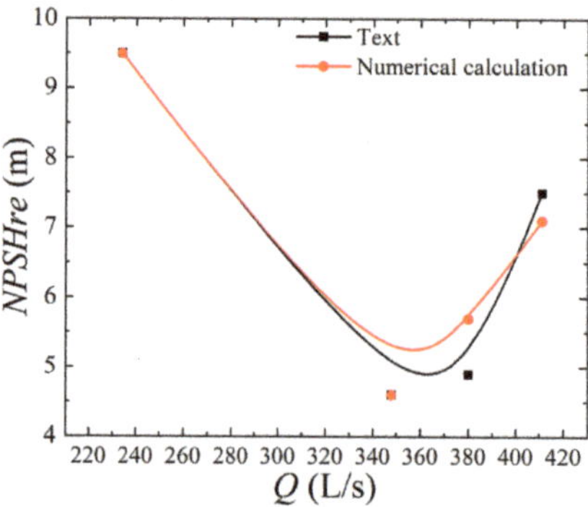

Figure 8. Comparison of the numerical calculation and the test critical cavitation margin.

According to Figures 7 and 8 and Table 4, the critical cavitation margin is 7.5 m and 4.9 m, respectively, under high flow conditions Q = 411 L/s and 380 L/s in the test. When Q = 348 L/s near the design condition, the critical cavitation margin is 4.6 m and reaches the minimum. Under low flow condition Q = 234 L/s, the critical cavitation margin is 9.5 m. The critical cavitation margin is 7.1 m, 5.7 m, 4.6 m, and 9.5 m for Q = 411 L/s, 380 L/s, 348 L/s, and 234 L/s, respectively, in the numerical calculation. $NPSH_{re}$ is required to be less than 5.5 m for the design condition (Q = 348 L/s), and the $NPSH_{re}$ is 4.6 m for both the numerical calculation and test results, which meets the design requirement of cavitation. Both the numerical calculation and test show that with the increase in flow rate, the critical cavitation margin first decreases, and the critical cavitation margin is the smallest near the high-efficiency zone. With the increase in flow rate, the critical cavitation margin continues to increase. The numerical calculation is close to the test critical cavitation margin, with small error, and shows essentially the same trend and high reliability of the numerical calculation.

Table 4. The numerical calculation and test critical cavitation error analysis table.

Q	$NPSH_{re}$ of Numerical Calculation	$NPSH_{re}$ of Test	Design Requirements	Calculation and Test Error
(L/s)	(m)	(m)	(m)	(m)
411	7.1	7.5	/	−0.4
380	5.7	4.9	/	0.8
348	4.6	4.6	<5.5	0
234	9.5	9.5	/	0

5. Analysis of Cavitation Numerical Calculation Results

Air bubble (caused by the release of non-condensable gases dissolved in the liquid due to a drop in pressure) cloud diagrams at different flow rates with the inlet pressure p_{in} = 11 × 10^4 Pa are organized as shown in Figure 9.

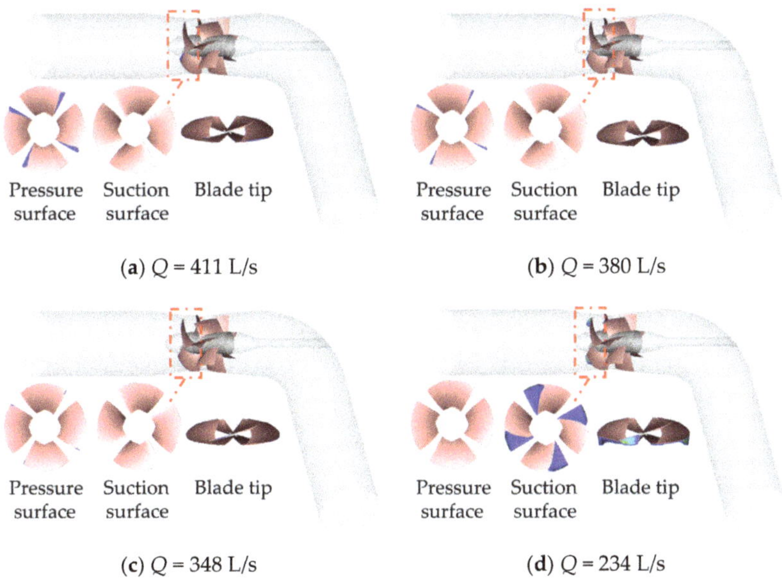

Figure 9. Air bubble distribution volume cloud diagrams at p_{in} = 11 × 10^4 Pa.

The comparative analysis of the cavitation at different flow rates with the same inlet pressure (p_{in} = 11 × 10^4 Pa) shows that the cavitation occurs at the inlet of the pressure surface under the design condition and high flow rate (Q = 348–411 L/s), presenting a strip-like distribution, while the cavitation occurs at the suction surface under the low flow rate (Q = 234 L/s), near the inlet blade tip, presenting a sheet-like distribution. As the flow rate decreases, the cavitation area of the pressure surface gradually decreases, and the area of the cavitation at the hub toward the wheel rim also gradually decreases; no air bubble appears at the pressure surface at the flow rate of Q = 234 L/s, at the suction surface, no air bubble appears at the high flow rate and design conditions (Q = 348–411 L/s), and the area of the air bubble is larger at the small flow rate conditions (Q = 234 L/s) and exceeds the most intense occurrence of cavitation when at the pressure surface. At the top of the leaf, there are vacuoles in all operating conditions; the vacuole area is small in the high flow condition and design condition (Q = 348–411 L/s), and the cavitation at the top of the leaf is more intense in the small flow condition (Q = 234 L/s), having a larger vacuole area and showing a cloud-like distribution.

The cavitation cloud diagrams at an inlet pressure of $p_{in} = 10 \times 10^4$ Pa and different flow rates are organized as shown in Figure 10.

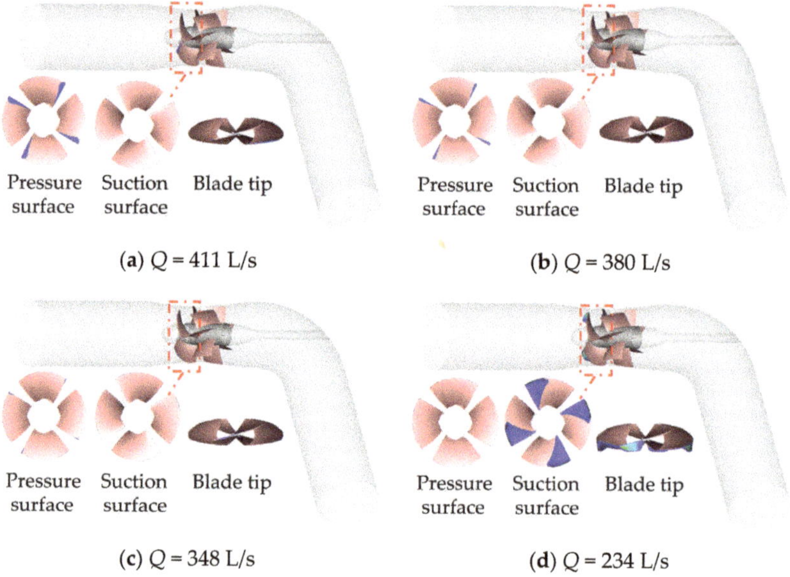

Figure 10. Air bubble distribution volume cloud diagrams at $p_{in} = 10 \times 10^4$ Pa.

The inlet pressure $p_{in} = 10 \times 10^4$ Pa, compared with the inlet pressure $p_{in} = 11 \times 10^4$ Pa, shows the same law of change with the working conditions. In the same working condition with the reduction in inlet pressure (inlet pressure from $p_{in} = 11 \times 10^4$ Pa to $p_{in} = 10 \times 10^4$ Pa), the area of the cavitation bubble increased and appeared in the same position. Because the inlet pressure $p_{in} = 10 \times 10^4$ Pa and the inlet pressure $p_{in} = 11 \times 10^4$ Pa show only small changes in the inlet pressure, and are born in the cavitation stage, the development of the degree of cavitation compared to the pressure change is not significant.

The cloud diagrams of cavitation at different flow rates with an inlet pressure $p_{in} = 8 \times 10^4$ Pa are organized as shown in Figure 11.

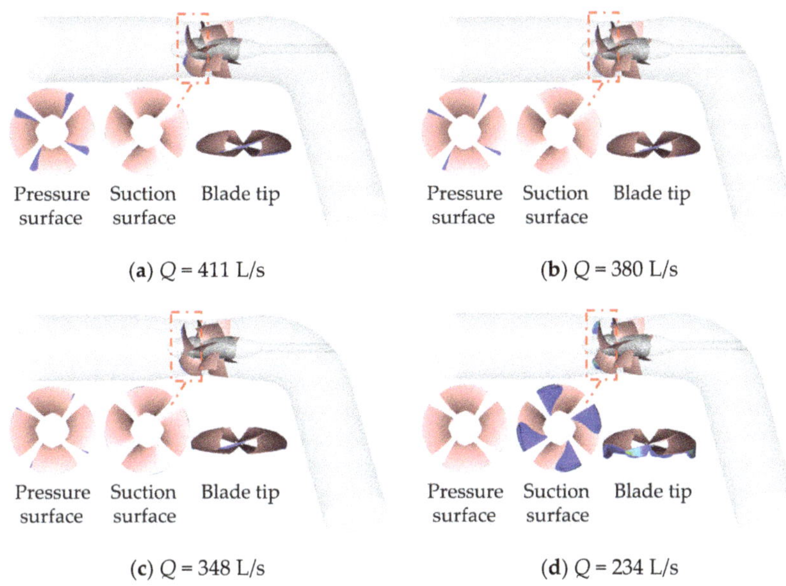

Figure 11. Air bubble distribution volume cloud diagrams at $p_{in} = 8 \times 10^4$ Pa.

The inlet pressure $p_{in} = 8 \times 10^4$ Pa, compared with the inlet pressures $p_{in} = 10 \times 10^4$ Pa and $p_{in} = 11 \times 10^4$ Pa, shows the law of change with the working condition is similar. In the same working condition, inlet pressure decreases (from $p_{in} = 10 \times 10^4$ Pa to $p_{in} = 8 \times 10^4$ Pa), the area of the air bubble further increases in comparison, and it can be seen that the area of the air bubble on the pressure surface increases significantly under the large flow rate and design working condition (Q = 348–411 L/s). Additionally, the cavitation area of the pressure surface increased noticeably, the cavitation area at the pressure surface still did not appear under the small flow condition (Q = 234 L/s), and the cavitation area at the suction surface still did not appear under the high flow and design conditions (Q = 348–411 L/s). The cavitation area at the blade tip increased obviously under the small flow condition (Q = 234 L/s), the cavitation area at the blade tip increased under each condition (Q = 234–411 L/s), and the cavitation area at the blade tip increased significantly under all operating conditions (Q = 234–411 L/s).

The cloud diagrams of cavitation at different flow rates with an inlet pressure of $p_{in} = 7 \times 10^4$ Pa are organized as shown in Figure 12.

The inlet pressure $p_{in} = 7 \times 10^4$ Pa, compared with the inlet pressure $p_{in} = 8 \times 10^4$ Pa, $p_{in} = 10 \times 10^4$ Pa, and $p_{in} = 11 \times 10^4$ Pa, shows the change law with the working condition is similar. In the same working condition with the decrease in inlet pressure (inlet pressure decreased from $p_{in} = 8 \times 10^4$ Pa to $p_{in} = 7 \times 10^4$ Pa), the area of the vacuole further increased in comparison, and the pressure surface change law was similar to the previous one. The vacuole appeared at the hub at the suction surface at the high flow condition (Q = 411 L/s) and the design condition (Q = 348 L/s), and the vacuole appeared at the suction surface, distributed in the middle of the blade head region. The cavitation at the blade tip at each working condition (Q = 234–411 L/s) is more obvious and starts to break away from the blade surface and develop into the fluid.

The air bubble cloud diagrams at different flow rates with an inlet pressure of $p_{in} = 6 \times 10^4$ Pa are organized as shown in Figure 13.

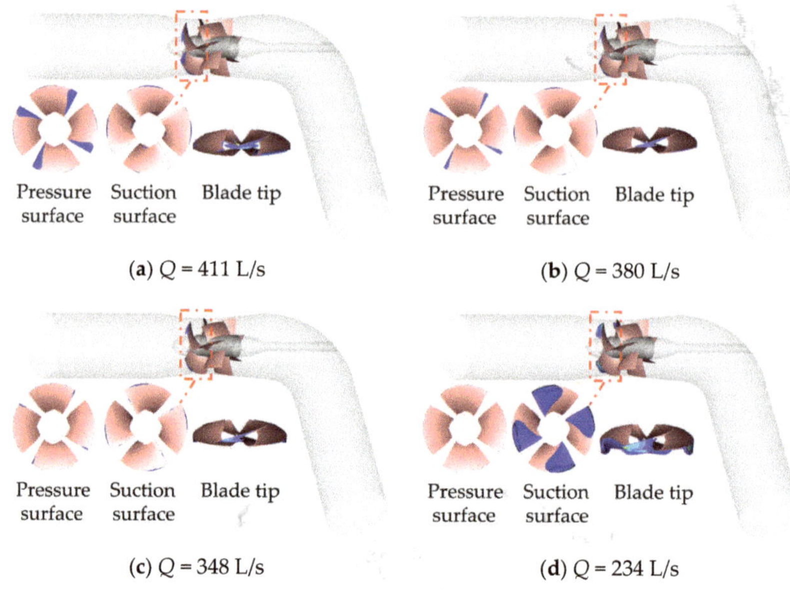

Figure 12. Air bubble distribution volume cloud diagrams at p_{in} = 7 × 10^4 Pa.

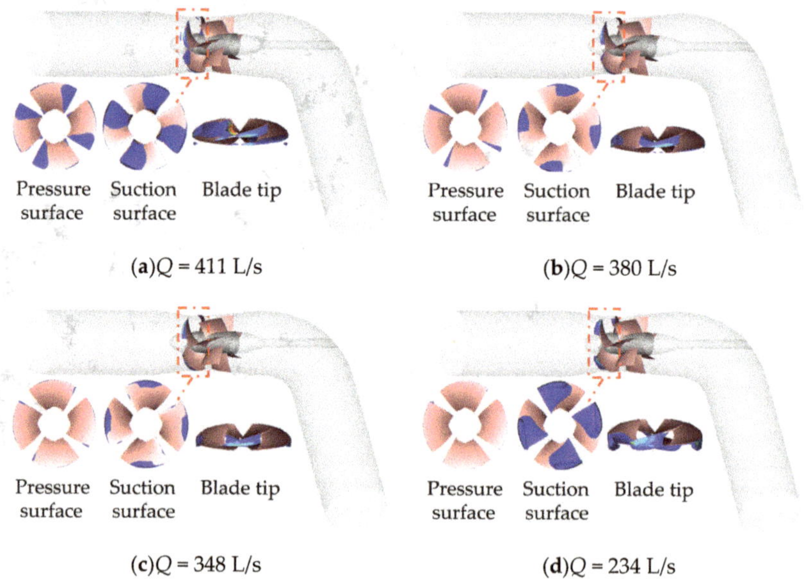

Figure 13. Air bubble distribution volume cloud diagrams at p_{in} = 6 × 10^4 Pa.

At an inlet pressure of p_{in} = 6 × 10^4 Pa, with the change of working conditions, the pressure surface of the air bubble area gradually reduced; the suction surface of the air bubble area was first reduced in the design working conditions (Q = 348 L/s) under the smallest air bubble area and then increased, but both the pressure surface or suction surface of each working conditions contained an air bubble.

In the same working condition as the inlet pressure decreases (inlet pressure decreased from p_{in} = 7 × 10^4 Pa to p_{in} = 6 × 10^4 Pa), the area of the vacuole further increases in comparison. The area of the vacuole at the hub further increases at the suction surface at the high flow condition (Q = 411 L/s), and the vacuole also exists at the suction surface at the design condition (Q = 348 L/s). The area of the vacuole distributed at the blade head in the middle region further increases, and the cavitation at the blade tip under each working condition (Q = 234–411 L/s) is more obvious as the vacuole produces obvious stripping and movement into the fluid.

The air bubble cloud diagrams at different flow rates with an inlet pressure of $p_{in} = 5 \times 10^4$ Pa are organized as shown in Figure 14.

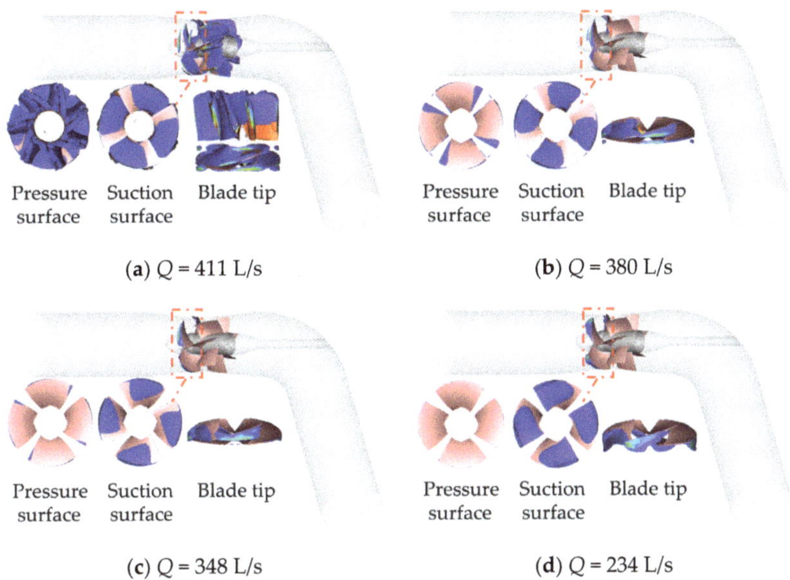

Figure 14. Air bubble distribution volume cloud diagrams at $p_{in} = 5 \times 10^4$ Pa.

The inlet pressure $p_{in} = 5 \times 10^4$ Pa, compared with the inlet pressure $p_{in} = 6 \times 10^4$ Pa, shows the same law of changing with the working conditions. In the same working condition, as the inlet pressure decreases (inlet pressure decreases from $p_{in} = 6 \times 10^4$ Pa to $p_{in} = 5 \times 10^4$ Pa), the area of the air bubbles further increases in comparison to the pressure surface in the high flow condition (Q = 411 L/s), the impeller and guide vane domain are basically full of air bubbles and the guide vane is also surrounded by air bubbles. The suction surface in the high flow condition (Q = 411 L/s) shows there is a large number of air bubbles at the hub; the air bubbles essentially wrapped 2/3 of the blade surface, in the design condition (Q = 348 L/s), the suction surface existing air bubbles essentially wrapped 1/2 of the blade surface and the cavitation at the top of the blade under each condition (Q = 234–411 L/s) is more obvious.

The air bubble cloud diagrams at different flow rates with an inlet pressure of $p_{in} = 4 \times 10^4$ Pa are organized as shown in Figure 15.

The inlet pressure $p_{in} = 4 \times 10^4$ Pa, compared with the inlet pressure $p_{in} = 5 \times 10^4$ Pa, shows the same law of changing with the working condition. The area of air bubbles further increases with the decrease in inlet pressure (from $p_{in} = 5 \times 10^4$ Pa to $p_{in} = 4 \times 10^4$ Pa) in the same working condition, and the pressure surface, suction surface, guide vane blade, and outlet bend are basically full of air bubbles under the high flow condition (Q = 411 L/s). At the design condition (Q = 348 L/s), the air bubbles existing at the suction surface basically wrap 4/5 of the blade surface. The cavitation at the top of the blade under each working condition (Q = 234–411 L/s) is more obvious.

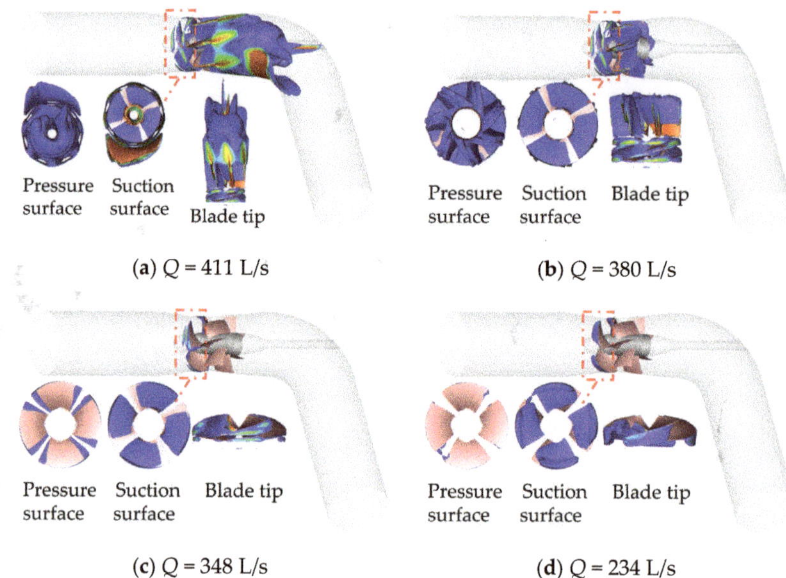

(**a**) $Q = 411$ L/s (**b**) $Q = 380$ L/s

(**c**) $Q = 348$ L/s (**d**) $Q = 234$ L/s

Figure 15. Air bubble distribution volume cloud diagrams at $p_{in} = 4 \times 10^4$ Pa.

The initial cavitation and critical cavitation pressure clouds at different flow rates were collated as shown in Figures 16–19.

Pressure: 52,950 56,983 107,025 110,843 117,438 120,820 125,825 130,814 134,103 137,181 140,800 144,332

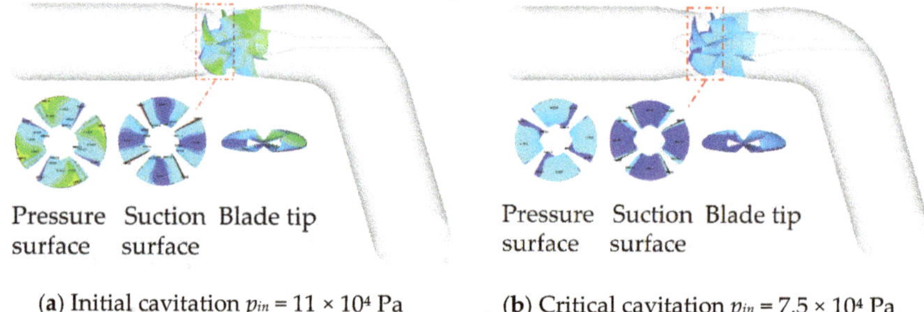

(**a**) Initial cavitation $p_{in} = 11 \times 10^4$ Pa (**b**) Critical cavitation $p_{in} = 7.5 \times 10^4$ Pa

Figure 16. Flow condition Q = 411 L/s pressure distribution cloud diagrams.

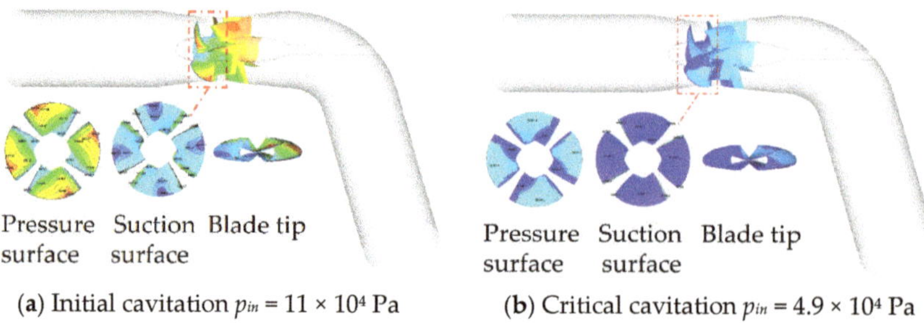

(**a**) Initial cavitation $p_{in} = 11 \times 10^4$ Pa (**b**) Critical cavitation $p_{in} = 4.9 \times 10^4$ Pa

Figure 17. Flow condition Q = 380 L/s pressure distribution cloud diagrams.

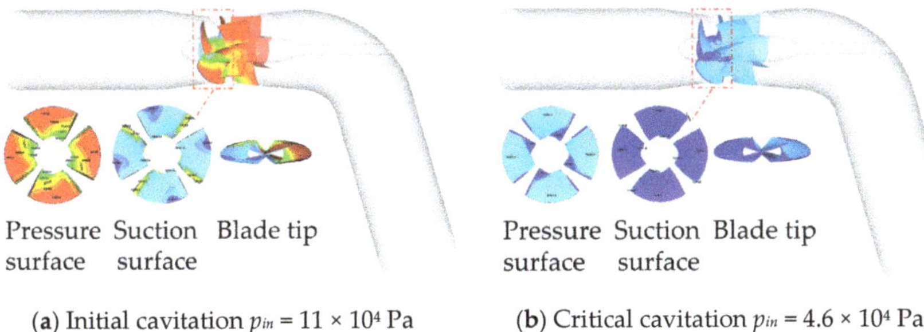

(**a**) Initial cavitation p_{in} = 11 × 10⁴ Pa (**b**) Critical cavitation p_{in} = 4.6 × 10⁴ Pa

Figure 18. Flow condition Q = 348 L/s pressure distribution cloud diagrams.

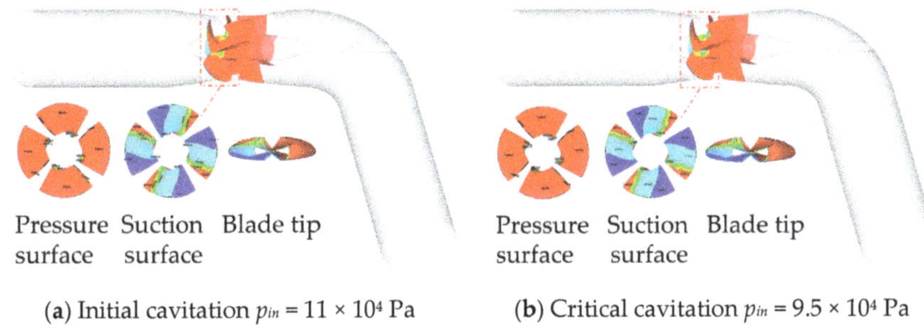

(**a**) Initial cavitation p_{in} = 11 × 10⁴ Pa (**b**) Critical cavitation p_{in} = 9.5 × 10⁴ Pa

Figure 19. Flow condition Q = 234 L/s pressure distribution cloud diagrams.

Through the comparative analysis of the pressure distribution cloud pictures at the suction surface, pressure surface, and blade tip in the same cavitation state under different flow conditions, it can be concluded that in the incipient cavitation state, with the increase in flow (Q = 234–411 L/s), the pressure on the surface gradually decreases, showing a gradient change from the hub to the rim, and the low-pressure area at the blade head gradually increases. The area of low pressure at the suction surface and the rim decreases first and then increases, the low-pressure area is then distributed at the blade head and the low-pressure area of the suction surface is distributed in a belt shape under the large flow condition (Q = 411 L/s) and in the small flow condition (Q = 234 L/s), the low-pressure area of the suction surface is distributed in a block shape near the design condition (Q = 348–380 L/s). In the critical cavitation state, with the increase in the flow (Q = 234–411 L/s), the low-pressure area on the pressure surface increases first and then decreases; the area of low pressure on the suction surface is smaller under the design condition and the small flow condition (Q = 234–348 L/s) and the area of low-pressure on the suction surface also increases first and then decreases. The low-pressure area is slightly smaller under the large flow condition (Q = 411 L/s) and the small flow condition (Q = 234 L/s), but the low-pressure area on the suction surface is also filled with more than 1/2 of the blade. This is mainly because when critical cavitation occurs, the inlet pressure under the large flow condition and small flow condition is large, while near the design condition (Q = 348–380 L/s), the cavitation performance is good, and the inlet pressure is small when cavitation occurs.

Through the comparative analysis of the pressure distribution clouds at the suction surface, pressure surface, and blade tip under different cavitation conditions under the same flow conditions, it can be concluded that the pressure at the pressure surface, suction surface, and blade tip during critical cavitation is less than that at the incipient cavitation. Near the large flow and design conditions (Q = 348–411 L/s), the low-pressure area at the blade head under critical cavitation is significantly larger than that under incipient cavitation. Under the small flow condition (Q = 234 L/s), the low-pressure area of the blade head does not increase significantly. Low pressure is the main inducing factor of cavitation, and the location of a low-pressure area can reflect the location and development of cavitation to a certain extent.

6. Conclusions

The numerical calculation and experimental analysis of the cavitation performance of the axial flow pump led to the following conclusions:

(1) In this paper, the numerical calculation of energy and cavitation performance of a high ratio speed axial flow pump based on the SST k-ω turbulence model and ZGB cavitation model is compared with the experimental analysis, the basic error of the head is within 0.2 m, the basic error range of efficiency is within 3%, and the error of critical cavitation margin is about 5%. Additionally, the numerical calculation trend is the same as the experimental, and, in general, the numerical calculation is better and more accurate at predicting the energy and cavitation performance of the axial flow pump.

(2) In the numerical simulation and experimental flow rate of about 350 L/s, the head is about 5.5 m and the efficiency is about 84.0%. The efficiency near the design point flow rate is high and the efficiency range in the high-efficiency zone is wide, which shows the reasonableness of the design. The maximum operating head is H_m = 8.80 m, which is 1.58 times the design head, giving the pump a wide operating range.

(3) In the test, the critical cavitation margin is 7.5 m and 4.9 m for the high flow conditions, Q = 411 L/s and 380 L/s, respectively, 4.6 m for Q = 348 L/s near the design condition, and 9.5 m for the low flow condition, Q = 234 L/s. In the numerical calculation, the critical cavitation margin is 7.1 m, 5.7 m, 4.6 m, and 9.5 m for Q = 411 L/s, 380 L/s, 348 L/s, and 234 L/s, respectively, and the error of critical cavitation margin is −0.4 m, 0.8 m, 0.0 m, and 0.0 m for each condition from high flow rate to low flow rate in numerical calculations and testing. The numerical calculations and experimental results follow the same trend with small errors. The $NPSH_{re}$ is less than 5.5 m at the design condition (Q = 348 L/s), and the numerical calculation and test result of $NPSH_{re}$ are 4.6 m, which meets the design requirement of cavitation. Both the numerical calculations and tests show that as the flow rate increases, the critical cavitation margin decreases first, and near the high-efficiency zone, the critical cavitation margin is minimal and continues to increase as the flow rate increases. The cavitation performance and energy of the high-ratio axial flow pump studied in this paper are excellent, which can provide a reference for the design and development of high-ratio axial flow pumps and the research of cavitation performance.

(4) According to the analysis of the numerical calculation results, with the same inlet pressure and with the reduction in flow, the first air bubbles appear in the head of the pressure surface of the large flow conditions, the pressure surface of the small flow conditions no longer have air bubbles, and air bubbles appear in the head of the suction surface. As the inlet pressure decreases, the air bubbles at the same flow rate gradually increased until there was violent cavitation, the air bubbles filled the impeller and guide vane area, and even at the outlet elbow there was a large number of air bubbles; the air bubbles moved downstream with the water flow.

7. Suggestions

This paper's numerical calculations and test results of the critical cavitation margin are consistent with the trend derived from previous studies. The critical cavitation margin showed the best cavitation performance near the design conditions, non-design conditions are slightly worse, and the birth location of the vacuole is also similar, but the distribution of the vacuole in this paper has been fully explored and reveals its movement and change of form. This paper's numerical calculations and test comparison have a high degree of agreement, revealing the spatial distribution characteristics of the vacuole inside the high ratio speed axial flow pump. The cavitation performance and energy performance of the axial flow pump are two contradictory quantities. The energy and cavitation characteristics of the high ratio speed axial flow pump studied in this paper are excellent, which can provide a certain basis for the development and design of the high ratio speed axial flow pump and provide a reference for the study of the cavitation mechanism of the high ratio speed axial flow pump. In the future, the authors will need to further investigate the cavitation characteristics of the axial flow pump based on acoustic analysis, time domain, and frequency domain vibration analysis, and carry out the structural optimization of the axial flow pump in the hope of further improving its cavitation performance.

Author Contributions: Concept design, C.X. and C.Z.; Numerical calculation, T.Z. and A.F.; Experiment and data analysis, T.F. and Y.J.; and Manuscript writing, C.X. and C.Z. All authors have read and agreed to the published version of the manuscript.

Funding: This research was funded by Anhui Province Natural Science Funds for Youth Fund Project, grant number 2108085QE220. Key scientific research project of Universities in Anhui Province, grant number KJ2020A0103. Anhui Province Postdoctoral Researchers' Funding for Scientific Research Activities, grant number 2021B552. Anhui Agricultural University President's Fund, grant number 2019zd10. Stabilization and Introduction of Talents in Anhui Agricultural University Research Grant Program, grant number rc412008.

Data Availability Statement: Not applicable.

Conflicts of Interest: The authors declare no conflict of interest.

References

1. Rayleigh, L. On the pressure developed in a liquid during the collapse of a spherical cavity. *Philos. Mag. Ser.* **1917**, *6*, 94–98. [CrossRef]
2. Plesset Milton, S.; Chapman Richard, B. Collapse of an initially spherical vapour cavity in the neighbourhood of a solid boundary. *J. Fluid Mech.* **1971**, *47*, 283–290. [CrossRef]
3. Kubota, A.; Kato, H.; Yamaguchi, H. A new modelling of cavitating flows: A numerical study of unsteady cavitation on a hydrofoil section. *J. Fluid Mech.* **1992**, *240*, 59–96. [CrossRef]
4. Singhal Ashok, K.; Athavale Mahesh, M.; Li, H.; Jiang, Y. Mathematical Basis and Validation of the Full Cavitation Model. *J. Fluids Eng.* **2002**, *124*, 617–624. [CrossRef]
5. Zuo, Z.; Liu, S.; Liu, D.; Qin, D. Numerical predictions and stability analysis of cavitating draft tube vortices at high head in a model Francis turbine. *Sci. China (Technol. Sci.)* **2014**, *57*, 2106–2114. [CrossRef]
6. Wang, Y.; Liu, H.; Liu, D.; Yuan, S.; Wang, J.; Jiang, L. Application of the two-phase three-component computational model to predict cavitating flow in a centrifugal pump and its validation. *Comput. Fluids* **2016**, *131*, 142–150. [CrossRef]
7. Zhang, D.; Shi, L.; Zhao, R.; Shi, W.; Pan, Q.; van Esch, B.P.M. Study on unsteady tip leakage vortex cavitation in an axial-flow pump using an improved filter-based model. *J. Mech. Sci. Technol.* **2017**, *31*, 659–667. [CrossRef]
8. Li, W.; Li, E.; Shi, W.; Li, W.; Xu, X. Numerical Simulation of Cavitation Performance in Engine Cooling Water Pump Based on a Corrected Cavitation Model. *Processes* **2020**, *8*, 278. [CrossRef]
9. Karakas, E.S.; Tokgöz, N.; Watanabe, H.; Aureli, M.; Evrensel, C.A. Comparison of Transport Equation-Based Cavitation Models and Application to Industrial Pumps with Inducers. *J. Fluids Eng.* **2021**, *144*, 011201. [CrossRef]
10. Pei, J.; Osman, M.K.; Wang, W.; Appiah, D.; Yin, T.; Deng, Q. A Practical Method for Speeding up the Cavitation Prediction in an Industrial Double-Suction Centrifugal Pump. *Energies* **2019**, *12*, 2088. [CrossRef]
11. Li, X.; Yuan, S.; Pan, Z.; Yuan, J.; Fu, Y. Numerical simulation of leading edge cavitation within the whole flow passage of a centrifugal pump. *Sci. China Technol. Sci.* **2013**, *56*, 2156–2162. [CrossRef]
12. Al-Obaidi, A.R. Effects of Different Turbulence Models on Three-Dimensional Unsteady Cavitating Flows in the Centrifugal Pump and Performance Prediction. *Int. J. Nonlinear Sci. Numer. Simul.* **2019**, *20*, 3–4.
13. Al-Obaidi, A.; Qubian, A. Effect of outlet impeller diameter on performance prediction of centrifugal pump under single-phase and cavitation flow conditions. *Int. J. Nonlinear Sci. Numer. Simul.* **2022**, *2020*, 0119. [CrossRef]
14. Feng, W.; Cheng, Q.; Guo, Z.; Qian, Z. Simulation of cavitation performance of an axial flow pump with inlet guide vanes. *Adv. Mech. Eng.* **2016**, *8*, 1687814016651583. [CrossRef]
15. Xu, B.; Shen, X.; Zhang, D.; Zhang, W. Experimental and Numerical Investigation on the Tip Leakage Vortex Cavitation in an Axial Flow Pump with Different Tip Clearances. *Processes* **2019**, *7*, 935. [CrossRef]
16. Al-Obaidi, A.R. Detection of Cavitation Phenomenon within a Centrifugal Pump Based on Vibration Analysis Technique in both Time and Frequency Domains. *Exp. Tech.* **2020**, *44*, 329–347. [CrossRef]
17. Al Obaidi, A.R. Experimental investigation of cavitation characteristics within a centrifugal pump based on acoustic analysis technique. *Int. J. Fluid Mech. Res.* **2020**, *47*, 501–515. [CrossRef]
18. Gao, B.; Guo, P.; Zhang, N.; Li, Z.; Yang, M. Experimental Investigation on Cavitating Flow Induced Vibration Characteristics of a Low Specific Speed Centrifugal Pump. *Shock. Vib.* **2017**, *2017*, 6568930. [CrossRef]

19. Li, Y.; Chen, C.; Pei, J.; Wang, W.; Wu, T. Pressure pulsation test of axial flow pump device under different cavitation conditions. *J. Agric. Mach.* **2018**, *49*, 158–164.
20. Shervani-Tabar, M.T.; Ettefagh, M.M.; Lotfan, S.; Safarzadeh, H. Cavitation intensity monitoring in an axial flow pump based on vibration signals using multi-class support vector machine. *Proc. Inst. Mech. Eng. Part C J. Mech. Eng. Sci.* **2018**, *232*, 3013–3026. [CrossRef]
21. Zhang, S.; Tian, R.; Ding, K.; Chen, H.; Ma, Z. Numerical and experimental study in pressure pulsation and vibration of a two-stage centrifugal pump under cavitating condition. *Mod. Phys. Lett. B* **2022**, *36*, 2150501. [CrossRef]
22. Tao, R.; Xiao, R.; Wang, F.; Liu, W. Improving the cavitation inception performance of a reversible pump-turbine in pump mode by blade profile redesign: Design concept, method and applications. *Renew. Energy* **2019**, *133*, 325–342. [CrossRef]
23. Xu, W.; He, X.; Hou, X.; Huang, Z.; Wang, W. Influence of wall roughness on cavitation performance of centrifugal pump. *J. Braz. Soc. Mech. Sci. Eng.* **2021**, *43*, 314. [CrossRef]
24. Al Obaidi, A.R. Numerical investigation on effect of various pump rotational speeds on performance of centrifugal pump based on CFD analysis technique. *Int. J. Modeling Simul. Sci. Comput.* **2021**, *12*, 2150045. [CrossRef]
25. Bae, J.H.; Chang, K.; Lee, G.H.; Kim, B.C. Bayesian Inference of Cavitation Model Coefficients and Uncertainty Quantification of a Venturi Flow Simulation. *Energies* **2022**, *15*, 4204. [CrossRef]
26. Xie, C.; Fu, T.; Xuan, W.; Bai, C.; Wu, L. Optimization and Internal Flow Analysis of Inlet and Outlet Horn of Integrated Pump Gate. *Processes* **2022**, *10*, 1753. [CrossRef]
27. Xie, C.; Yuan, Z.; Feng, A.; Wang, Z.; Wu, L. Energy Characteristics and Internal Flow Field Analysis of Centrifugal Prefabricated Pumping Station with Two Pumps in Operation. *Water* **2022**, *14*, 2705. [CrossRef]
28. Xie, C.; Xuan, W.; Feng, A.; Sun, F. Analysis of Hydraulic Performance and Flow Characteristics of Inlet and Outlet Channels of Integrated Pump Gate. *Water* **2022**, *14*, 2747. [CrossRef]
29. Shi, L.; Tang, F.; Zhou, H.; Tu, L.; Xie, R. Axial-flow pump hydraulic analysis and experiment under different swept-angles of guide vane. *Editor. Off. Trans. Chin. Soc. Agric. Eng.* **2015**, *31*, 90–95.
30. Zhang, W.; Tang, F.; Shi, L.; Hu, Q.; Zhou, Y. Effects of an Inlet Vortex on the Performance of an Axial-Flow Pump. *Energies* **2020**, *13*, 2854. [CrossRef]
31. Shen, S.; Huang, B.; Huang, S.; Xu, S.; Liu, S. Research on Cavitation Flow Dynamics and Entropy Generation Analysis in an Axial Flow Pump. *J. Sens.* **2022**, *2022*, 7087679. [CrossRef]
32. Yang, F.; Zhao, H.R.; Liu, C. Improvement of the Efficiency of the Axial-Flow Pump at Part Loads due to Installing Outlet Guide Vanes Mechanism. *Math. Probl. Eng.* **2016**, *2016*, 6375314. [CrossRef]

 water

MDPI

Article

Prediction for the Influence of Guide Vane Opening on the Radial Clearance Sediment Erosion of Runner in a Francis Turbine

Zhiqiang Jin [1], Xijie Song [2], Anfu Zhang [1], Feng Shao [1] and Zhengwei Wang [2,*]

1 Xinjiang Xinhua Hydroelectric Investment Technology Co., Ltd., Akesu 843300, China
2 State Key Laboratory of Hydroscience and Engineering, Department of Energy and Power Engineering, Tsinghua University, Beijing 100084, China
* Correspondence: wzw@mail.tsinghua.edu.cn

Abstract: In this paper, the Eulerian–Lagrangian method and Tabakoff erosion model are used to study the solid–liquid two-phase flow in a Francis turbine. Through the analysis of the overall flow pattern, particle flow, particle concentration, and wear in the bladeless area of the unit under different guide vane openings, the influence of runner radial gap flow on the surrounding flow field characteristics and wear under different guide vane openings is revealed. The results show that the smaller the opening of the guide vane, the greater the influence on the vortices and flow pattern and the particle distribution in the runner. The overall wear in the hydraulic turbine unit with the optimal opening is the smallest. The long-term wear of the runner inlet and guide vane outlet will cause the loss of local structures, an increase in the radial clearance of the runner, an increase in the clearance leakage, an increase in the vibration of the unit, and a reduction in efficiency. The research results provide a basis for the structural and hydraulic optimization of the Francis turbine.

Keywords: Francis turbine; guide vane opening; sediment erosion; clearance; CFD

Citation: Jin, Z.; Song, X.; Zhang, A.; Shao, F.; Wang, Z. Prediction for the Influence of Guide Vane Opening on the Radial Clearance Sediment Erosion of Runner in a Francis Turbine. *Water* **2022**, *14*, 3268. https://doi.org/10.3390/w14203268

Academic Editor: Giuseppe Pezzinga

Received: 16 September 2022
Accepted: 13 October 2022
Published: 17 October 2022

Publisher's Note: MDPI stays neutral with regard to jurisdictional claims in published maps and institutional affiliations.

Copyright: © 2022 by the authors. Licensee MDPI, Basel, Switzerland. This article is an open access article distributed under the terms and conditions of the Creative Commons Attribution (CC BY) license (https://creativecommons.org/licenses/by/4.0/).

1. Introduction

China's rivers contain a lot of sand; moreover, 115 rivers have an average annual sediment discharge of more than 10 million tons. Regarding the Yangtze River, the average annual sediment discharge has reached 514 million tons; the average annual sediment concentration in the Three Gorges Reservoir Area is 1.17 kg/m^3 and the maximum is 10.5 kg/m^3. The sediment concentration in the Yellow River is higher. According to statistics, the average annual sediment concentration of the Sanmenxia reach of the Yellow River is 37.6 kg/m^3. The high sediment concentration in rivers in China leads to sediment abrasion of 30~40% of hydropower stations [1,2].

Hydraulic machinery has therefore been operating in the water flow with high sediment volume for a long time, which has caused different degrees of abrasion of the turbines of most power stations in China [2]. Severe wear causes damage to the structural materials of the hydropower equipment itself, affects the reliability and stability of the operation, reduces the efficiency and output of the turbine, shortens the service life of the runner, prolongs the construction period, increases the consumption of materials and spare parts, and causes huge economic losses [3,4].

During the rotation process of the Francis turbine, the pressure fluctuation in the non-blade area will be caused by the comprehensive influence of the flow field distortion caused by the runner blades and guide vanes [5]. The radial clearance of the runner is the only channel connecting the vaneless area with the lower cavity of the top cover and the upper cavity of the bottom ring, so the hydraulic characteristics in the vaneless area have a certain correlation with the hydraulic characteristics of the lower cavity of the top cover and the upper cavity of the bottom ring [6]. The wear in the clearance will change the clearance size, and the change in the clearance size will have an impact on the flow field characteristics around the runner [7]. The change in flow field characteristics around the runner may lead to the increase in unit vibration and power swing, and even threaten the safe and stable operation of the power station [8].

Wear is the process of material transfer and loss in the contact surface layer during the relative movement of interacting solid surfaces. The working head of the hydraulic turbine is high, and the internal wear is mainly erosion wear [9]. In order to reveal the erosion wear mechanism, many researchers have studied the erosion wear from experimental and theoretical aspects, and put forward a variety of erosion wear prediction models. Finnie proposed the first erosion model. The model assumes that particles do not break up in the process of cutting metal, and considers that the erosion wear of solid particles on ductile materials is mainly due to the cutting effect of particles on materials [10]. Tabakoff introduces the impact angle and impact velocity to the base Finnie model, which has good prediction ability for the erosion and wear in the hydraulic turbine and has been widely used.

Sediment erosion mainly belongs to sediment-laden flow, which is a typical solid–liquid two-phase flow [11]. At present, a lot of studies have been performed on the Francis turbine, from the aspects of structure design, design method, basic equation, flow law, and wear mechanism. Li [12] studied the flow field characteristics of turbine guide vanes under the condition of sediment-laden flow, and found the best relative placement position of guide vanes and fixed guide vanes under the condition of sediment-laden flow. Liu [13] established the Euler–Lagrangian mixed turbulence model of low concentration solid–liquid two-phase flow, and produced the particle wall collision model and the erosion model of hydraulic turbine flow passage parts made of ductile metal materials. These models can be used to numerically simulate the flow of water containing sand in the hydraulic turbine, the concentration distribution of sand particles in the flow passage of the hydraulic turbine, the movement track, and the erosion rate of hydraulic turbine flow passage parts. Qi [14] derived the energy equation and energy dissipation term expression of solid–liquid two-phase flow describing sediment-laden flow from the energy equation of continuous medium flow, and introduced the energy dissipation extreme value principle into the study of sediment-laden flow characteristics in hydraulic turbines. Li [15] reported that water turbines in sediment-laden flow are prone to abrasion and damage. For the built hydropower station, adopting a reasonable operation mode and certain maintenance measures can delay or reduce the abrasion of the water turbine. Huang [16] expounded the micro process and mechanism of abrasion damage. The stability is related to the safe and normal operation of the unit. The sediment-laden flow is different from the single-phase flow of clean water. It is also necessary to explore the stability of the unit in sediment-laden flow.

The research background of this paper is Xinjiang Tagake Hydropower Station, which is located 14 km from the head of the Xiehera diversion canal in the Aksu region, Xinjiang. It is a runoff diversion hydropower station, with a diversion canal of 6.88 km long and a tailrace of 4.22 km long. Two mixed flow turbine generator units with a unit capacity of 24.5 MW are installed. The design diversion flow is 75 m^3/s, the rated head is 74 m, the guaranteed output is 13.8 MW, and the annual design power generation is 273.9 million kW·h. The high sediment concentration in Xinjiang makes the unit have serious sediment abrasion problem.

Due to the lack of research on clearance wear of hydraulic turbine, in this paper, numerical simulation is used to predict the sediment-laden flow and wear of a hydraulic turbine unit in the power station. A full fluid domain model of real machine size is established and the influence of runner radial clearance and clearance wear on the flow field characteristics around the runner under different guide vane openings was simulated through CFX, so as to improve the mechanical and hydraulic performance of the unit.

2. Mathematical and Geometrical Models

2.1. Mathematical Model

2.1.1. Governing Equations

In this paper, the sediment laden flow is simulated and solved by Euler–Lagrange method. In the Euler–Lagrange system, the fluid phase is treated as a continuous phase, solved by Euler methods, and the particle phase as a discrete phase, solved by Lagrange methods [17]. The flow control equation of continuous phase is solved by the *N-S* equation.

$$\frac{\partial(\rho u)}{\partial t} + \nabla \bullet (\rho uu) = -\nabla p + \rho v \Delta u - \rho \nabla \bullet \tau + S_t \tag{1}$$

where u is the flow velocity, t is the time, ρ is the fluid density, p is the flow pressure, ν is the kinematic viscosity of the fluid, and S_t is the source term. τ is the Reynolds stress defined as:

$$\tau = \tau^d + \frac{2k}{3}\delta \tag{2}$$

where τ^d is the deviatoric Reynolds stress, k is the turbulent kinetic energy, δ is the Kronecker delta. Based on viscosity (ν_t) assumptions, Equation (2) can be written as:

$$\tau = -2v_t S + \frac{2k}{3}\delta \tag{3}$$

where S is the train-rate tensor,

$$S = \frac{1}{2}\left(\nabla u + \nabla^T u\right) \tag{4}$$

2.1.2. Lagrangian Tracking of Particle Motion

The particle movement in the hydraulic turbine is mainly represented as discrete phase movement, so the Lagrangian particle tracking model, which is widely used in sediment movement, is used to track the particle movement in the hydraulic turbine.

$$m_p \frac{du_p}{dt} = F_D + F_B + F_G + F_V + F_P + F_X \tag{5}$$

where t is time, m_p is particle mass, u_p is particle velocity, F_D is resistance, F_B is Basset force, F_G is gravity, F_V is virtual mass force, F_P is pressure gradient force, and F_X is the sum of other external forces considered.

In this paper, the particle concentration in the flow field is small, the fluid velocity of the continuous phase in the pump is large, and there is a large density difference between the continuous phase and the discrete phase. Therefore, the virtual mass force, pressure gradient force, Basset force, Saffman force, and Magnus force on the solid particles can be ignored. The basic equation of particle motion can be expressed as:

$$\frac{dx_{pi}}{dt} = u_{pi} \tag{6}$$

$$\frac{du_{pi}}{dt} = \frac{3C_{D\rho f}}{4\rho_p D_p}|u_s|u_s \tag{7}$$

where u_s is the slip velocity between particles and the liquid, C_D is the drag coefficient related to Reynolds number, ρ_f is the liquid density, ρ_p is the particle density, D_p is the particle diameter, and x_{pi} is the spatial coordinate position of particles.

In this paper, one-way coupling between the fluid phase and the solid phase is adopted in the particle track model considering the particle concentration.

2.1.3. Erosion Model

The internal sediment erosion of hydraulic turbine is mainly impact wear, and the commonly used impact erosion models include the Finnie erosion model, Tabakoff erosion model, and Oka Erosion model [18]. Among them, the Tabakoff erosion model is widely used in the prediction of mixed flow turbine sediment erosion in engineering [19,20].

Therefore, in this paper, the Tabakoff erosion model is used to predict the sediment erosion in the turbine.

$$E = k_1 f(\gamma) V_P^2 \cos^2\gamma \left[1 - R_T^2\right] + f(V_{PN}) \quad (8)$$

$$f(\gamma) = \left[1 + k_2 k_{12} \sin\left(\gamma \frac{\pi/2}{\gamma_0}\right)\right]^2 \quad (9)$$

$$R_T = 1 - k_4 V_P \sin\gamma$$

$$f(V_{PN}) = k_3 (V_P \sin\gamma)^4$$

$$k_2 = \begin{cases} 1.0 \text{ if } \gamma \le 2\gamma_0 \\ 0.0 \text{ if } \gamma > 2\gamma_0 \end{cases}$$

Here, γ_0 is the angle of maximum erosion, k_1 to k_4, $k_{12,}$ and γ_0 are model constants and depend on the particle/wall material combination.

2.2. *Simulation Geometry Model*

2.2.1. Geometric Model Set Up

Figure 1 shows the flow diagram in the runner of the hydraulic turbine. The clearance flow between the runner and the fixed parts is complex, which can easily cause sediment abrasion.

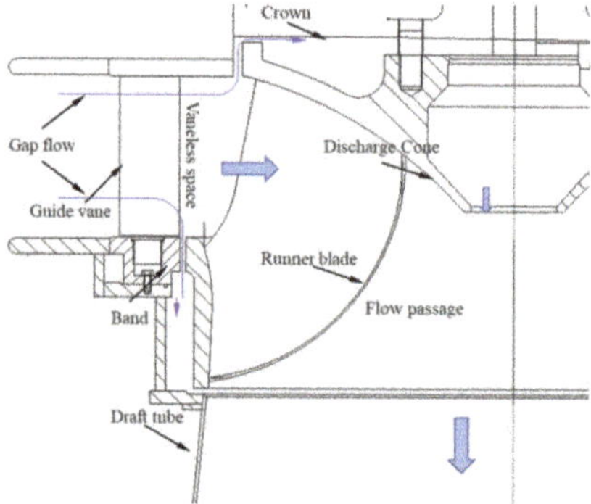

Figure 1. Flow diagram in runner.

A three-dimensional calculation model of the whole flow channel including volute, fixed guide vane, guide vane, runner, upper crown cavity, lower ring cavity, draft tube and outlet reservoir is established, as shown in the Figure 2.

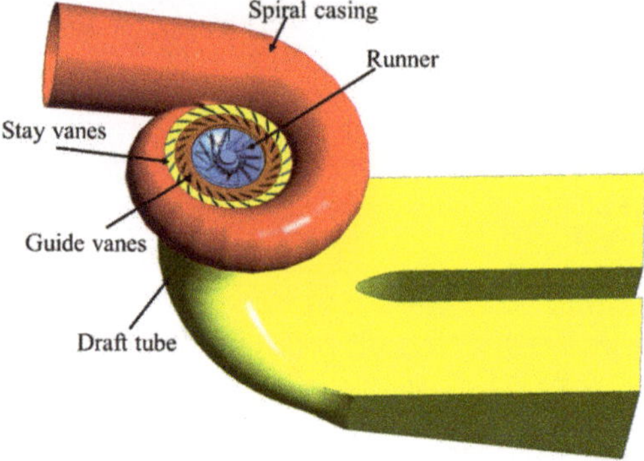

Figure 2. Model of numerical simulation.

The computational geometric model is meshed; a structured mesh is adopted for the gap and a hybrid mesh is adopted for other parts, as shown in Figure 3. The efficiency of hydraulic turbine is selected as the criterion for grid independence verification. Figure 4 clearly shows the relationship between turbine efficiency and grid number. When the number of grids increases from 11.2 million to 12.3 million, the absolute increment of turbine efficiency at the optimal operating point is less than 0.01%, so 11.2 million grids are used for the numerical calculation.

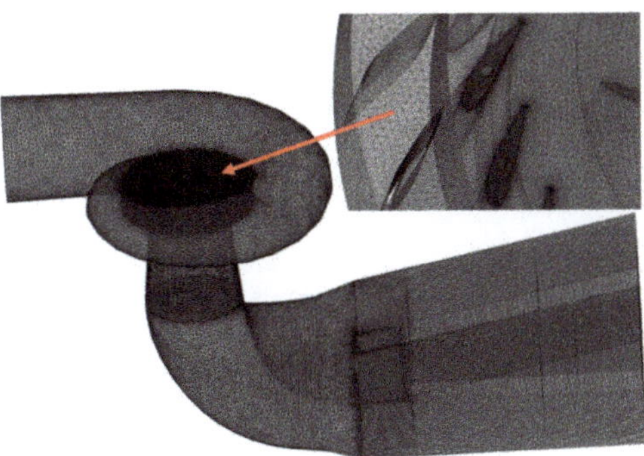

Figure 3. Clearance grids.

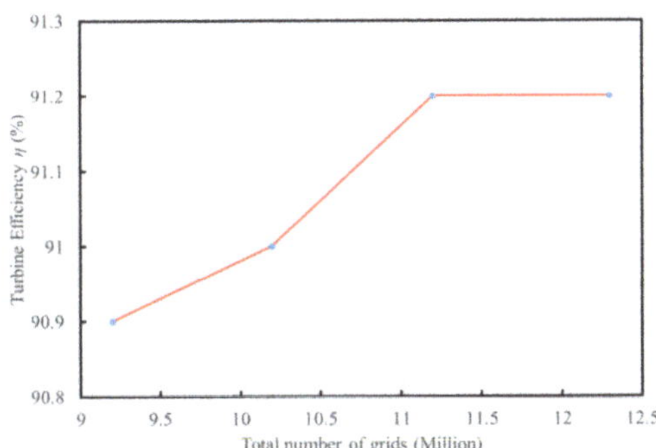

Figure 4. Efficiency of turbine with different grid numbers.

2.2.2. Parameter Setting in Calculation Model

(1) Boundary conditions

In this paper, the Ansys CFX software is used to simulate the flow of sand in the full channel of hydraulic turbines with different guide vane openings. The inlet boundary was set to the total pressure corresponding to the water head of the upstream, and the outlet boundary was adopted the static pressure condition related to the water level of downstream. The wall of the unit adopted the no-slip boundary. The interface between the runner and stationary parts adopts the dynamic–static interface.

(2) Calculation parameters

In the calculation scheme, the particle concentration is 10 kg/m^3, the guide vane openings are 15°, 20°, and 30°; among these values, 20° of opening is the optimal opening (see Figure 5). The solution precision is set to 10^{-5}. The interphase coupling between the fluid phase and the solid phase adopted the one-way coupling for the low particle concentration.

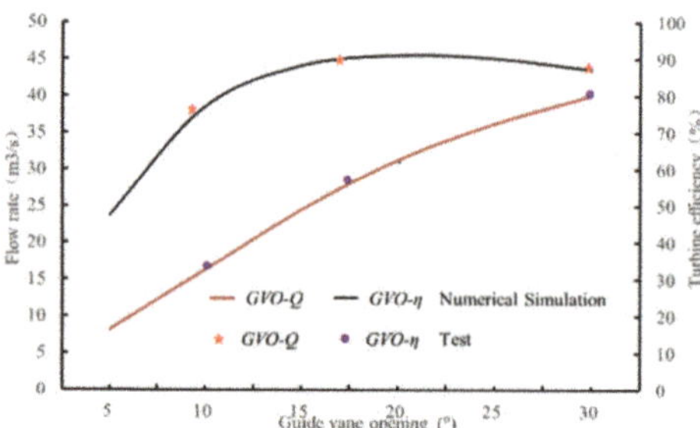

Figure 5. Reliability verification of external characteristics of hydraulic turbine.

2.3. Reliability Verification of Calculation Model

The reliability of the simulated results is verified by the field sediment erosion diagram of the runner of the hydraulic turbine unit and the field operation data of the unit. Figure 5 shows the calculated wear prediction and field wear comparison of the unit clearance and runner blades. The reliability of the calculation results is verified by the field wear diagram of the runner of the hydraulic turbine unit and the field operation data of the unit. The operation efficiency of the unit is obtained through steady calculation of the unit under different guide vane openings. Figure 6 shows the calculated wear prediction and field wear comparison of the unit clearance and runner blades. The clearance wear is consistent with the wear characteristics of water turbine on site. Figure 5 shows the operation curve of unit, and the tested data of unit operation is provided by Tagake Hydropower Station. The predicted results of operation characteristics and sediment erosion show that the numerical simulation method is reliable.

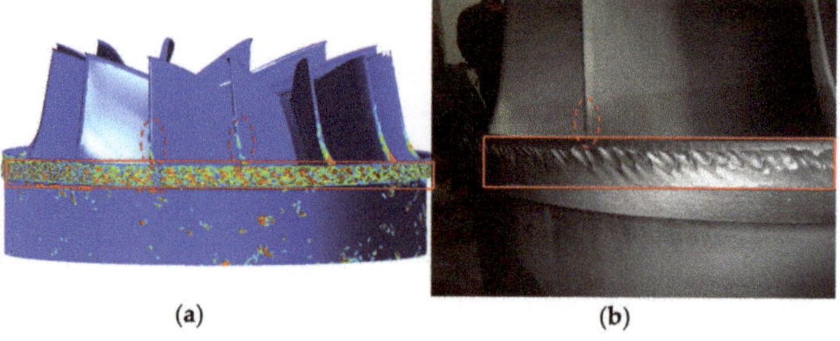

Figure 6. Sediment erosion at the clearance of the band. (**a**) CFD result; (**b**) physical wear picture.

3. Result Analysis

3.1. Flow Pattern of Turbine under Different Openings

Select the flow distribution on the central longitudinal section of the turbine to analyze the flow pattern in the turbine with different guide vane openings, as shown in Figure 7. When the guide vane opening is 15°, the flow pattern along the turbine is very poor, the flow velocity in the runner is low, and there are backflow and vortices in the draft tube. As shown in Figure 7a, these vortices and backflow increase the instability of the water flow and lead to the vibration of the unit. When the guide vane opening is 20°, the flow pattern in the turbine is obviously better than that in the turbine with the guide vane opening of 15°, as shown in Figure 7b. When the guide vane opening is 30°, the flow in the runner is very smooth, the flow velocity in the runner is the highest, and there is no backflow or vortices in the draft tube, as shown in Figure 7c. However, the increase in the flow velocity will also lead to more hydraulic loss and higher impact velocity of sand particles.

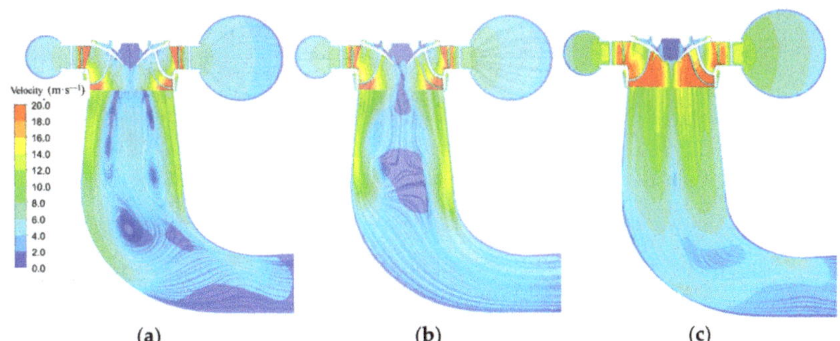

Figure 7. Overall flow pattern of internal section of hydraulic turbine with different opening. (**a**) 15° (**b**) 20° (**c**) 30°.

Figures 8–10 shows the flow pattern in the guide vane and the runner on the horizontal section passing through the center of the runner. It can be seen from the figure that due to the small opening, the velocity triangle of the water flow at the inlet of the blade head completely deviates from the design velocity triangle of the hydraulic turbine, the velocity circulation of the guide vane outlet cannot meet the velocity circulation requirements of the runner blade, and the relative velocity angle of the water flow entering the runner is larger than the inlet angle of the runner blade, resulting in the water flow hitting the pressure surface of the runner blade at a large angle. The flow water loses the constraint effect of the blade, forms a blade passage vortex in the runner, increases the hydraulic loss of the runner blade, and reduces the hydraulic power generation efficiency of the turbine.

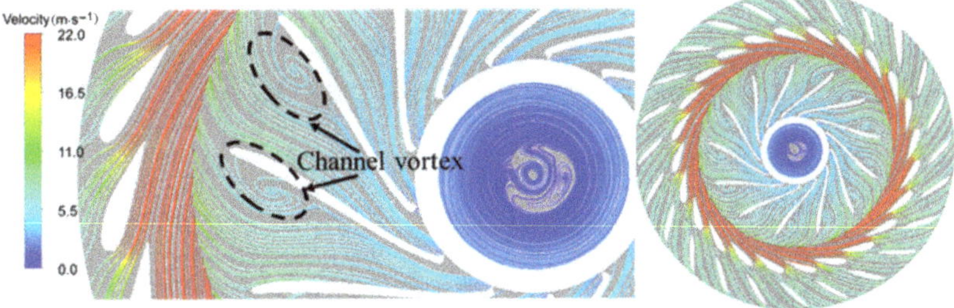

Figure 8. Flow pattern of vaneless space at 15 degrees of opening.

Figure 9. Flow pattern of vaneless space at 20° of opening.

Figure 10. Flow pattern of vaneless space zone at 30° of opening.

3.2. Sediment Distribution at Different Guide Vane Openings

In order to further explore the flow characteristics and sediment distribution characteristics in the vaneless area, the particle distribution and solid volume distribution number distribution in the local vaneless area are selected for analysis. The horizontal section passing through the center of the runner is as shown in Figure 11.

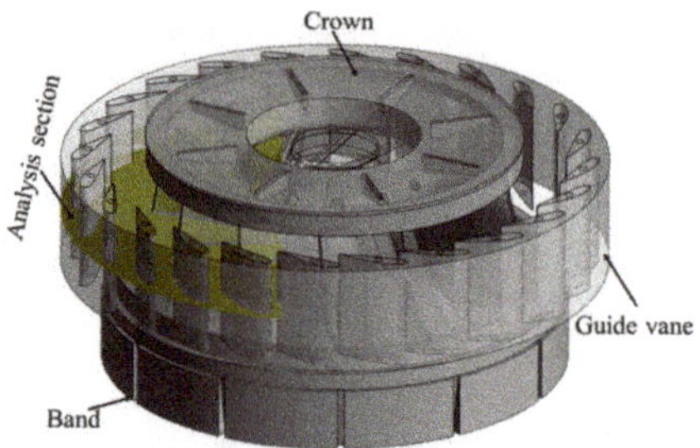

Figure 11. Schematic diagram of analysis section.

The sediment distribution in the guide vane and runner channel under different guide vane opening is shown in Figure 12. Figure 12 clearly shows that the sediment concentration in the guide vane channel and runner channel gradually decreases with the increase in the guide vane opening. When the opening of the guide vane is 15°, there is obvious sediment accumulation in the leafless area between the guide vane and the runner. With the increase in the opening of the guide vane, the sediment concentration in the channel of the guide vane gradually decreases, and the sediment is carried into the runner channel, which indicates that the increased opening of the guide vane will cause more wear and damage to the runner blade. However, with the increase in guide vane opening, the velocity of sediment particles increases, and the impact velocity on the guide vane and runner wall increases. Particles have more energy to destroy the wall structure of the runner and blade [21]. The smaller the opening, the higher the sediment concentration in the runner and guide vane channel. In the low-velocity region of the vortex return center, more particles are separated from the main flow into the vortex center of the channel and away from the runner wall [22,23], which is consistent with the flow change law in the runner and guide vane.

3.3. Sediment Erosion Distribution of Guide Vane and Runner Blade Wall

The sediment erosion distribution for the guide vane wall and runner blade wall under different guide vane openings is shown in Figure 13. There are different erosion rates on the guide vane and runner blade under different guide vane openings, and the wear is mainly distributed on the inlet head and outlet wall. When the guide vane opening of the runner is 20°, there is a little wear on the guide vane and the runner blade. Under this opening, the wear of the guide vane only exists at the head of the guide vane, and there is no wear on the outlet surface of the guide vane. When the guide vane opening is 15° and 30°, the sediment erosion on the guide vane and runner blade is clearly serious, and the range and erosion rate of impact erosion and friction erosion are greatly increased, as shown in Figure 13a,b. This shows that the overall wear in the hydraulic turbine unit with the optimal opening is the smallest, the erosion rate increases under the small opening and the large opening, the wear of the guide vane and the runner inlet head is mainly impact wear, and the wear on the outlet wall is mainly friction wear, and the wear in the unit channel is the most serious under the large opening [24]. The long-term wear of the runner inlet and guide vane outlet will cause the loss of local structure in the bladeless area, increase the radial clearance of the runner, increase the clearance leakage, increase the vibration of the unit, and reduce the efficiency [25,26].

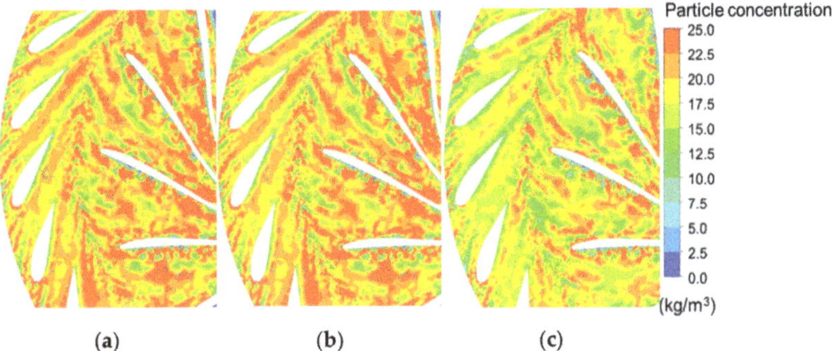

Figure 12. Volume fraction distribution of local solid phase in guide vane and runner passage: (**a**) 15°, (**b**) 20°, (**c**) 30°.

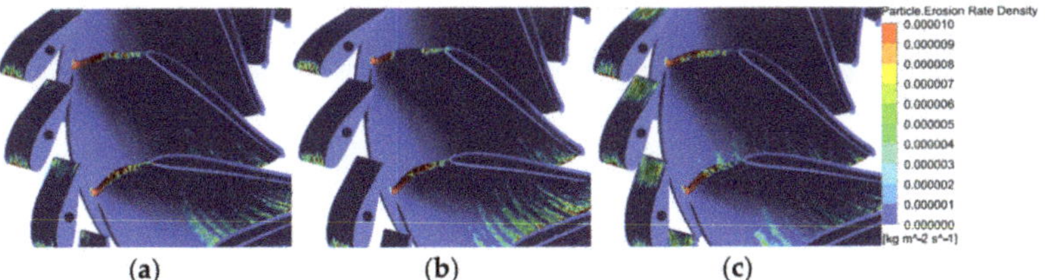

Figure 13. Sediment erosion distribution of guide blade and runner blade under different openings: (**a**) 15°, (**b**) 20°, (**c**) 30°.

4. Conclusions

This paper reveals the influence of radial clearance wear of a Francis pump turbine runner on the surrounding flow field characteristics under different guide vane openings, and obtains the following conclusions:

(1) The increase in guide vane opening has an important effect on the particle motion in the runner. With the increase in guide vane opening, the velocity of sediment particles increases, and the impact velocity on the guide vane and runner wall increases. The particles have more energy to destroy the wall structure of the runner and blade. The smaller the opening, the higher the sediment concentration in the runner and guide vane channel. In the low-velocity region of the vortex return center, more particles are separated from the main flow into the vortex center of the channel and away from the runner wall. This is consistent with the flow change law in the runner and guide vane.

(2) The opening of the guide vane affects the flow in the hydraulic turbine channel. When the opening of the guide vane is small, the velocity triangle of the water flow at the

inlet of the blade head completely deviates from the design velocity triangle of the hydraulic turbine, the velocity loop of the water flow out of the guide vane cannot meet the velocity loop required by the runner blade, and the relative velocity angle of the water flow entering the runner is larger than the inlet angle of the runner blade, resulting in the water flow hitting the pressure surface of the runner blade at a large angle, and the flow phenomenon occurs at the head of the runner blade. The separated flow loses the restriction of the blade and forms a blade passage vortex in the runner.

(3) The overall sediment erosion in the hydraulic turbine unit with the optimal opening is the smallest, and the erosion rate increases under a small opening and a large opening. The wear of the guide vane and the runner inlet head is mainly impact wear, and the sediment erosion on the outlet wall is mainly frictional wear. The sediment erosion in the unit channel is most serious under a large opening. The long-term wear of the runner inlet and guide vane outlet will cause the loss of local structure, increase the radial clearance of the runner, increase the clearance leakage, increase the vibration of the unit, and reduce the efficiency.

Author Contributions: Data curation, X.S.; software, validation, Z.J.; formal analysis, investigation, A.Z.; resources, F.S.; writing—original draft preparation, X.S.; writing—review and editing, Z.W. All authors have read and agreed to the published version of the manuscript.

Funding: This work was supported by a research project on sediment abrasion mechanism of turbine and R&D of anti-abrasion runners in Tagake Hydropower Station [XHTS-A-WZ-2022-005], National Natural Science Foundation of China (51876099).

Informed Consent Statement: Not applicable.

Data Availability Statement: Not applicable.

Conflicts of Interest: The authors declare no conflict of interest.

References

1. Kim, T.S.; Cha, K.S. Comparative analysis of the influence of labyrinth seal configuration on leakage behavior. *J. Mech. Sci. Technol.* **2009**, *23*, 2830–2838. [CrossRef]
2. Takaffoli, M.; Papini, M. Numerical simulation of solid particle impacts on Al6061-T6 part I: Three-dimensional representation of angular particles. *Wear* **2012**, *292*, 100–110. [CrossRef]
3. Song, X.; Liu, C.; Wang, Z. Prediction on the pressure pulsation induced by the free surface vortex based on experimental investigation and Biot-Saval Law. *Ocean Eng.* **2022**, *250*, 110934. [CrossRef]
4. Song, X.; Luo, Y.; Wang, Z. Numerical prediction of the influence of free surface vortex air- entrainment on pump unit performance. *Ocean Eng.* **2022**, *256*, 111503. [CrossRef]
5. Cheng, X.; Dong, F.; Yang, C. The influence of particle diameter on friction loss intensity and collision wear loss intensity along blades of centrifugal pump. *J. Lanzhou Univ. Technol.* **2015**, *41*, 44.
6. Song, X.; Liu, C. Experiment study of the floor-attached vortices in pump sump using V3V. *Renew. Energy* **2021**, *164*, 752–766. [CrossRef]
7. Matsumura, T.; Shirakashi, T.; Usui, E. Identification of Wear Characteristics in Tool Erosion model of Cutting Process. *Int. J. Mater. Form.* **2008**, *1*, 555–558. [CrossRef]
8. Koirala, R.; Thapa, B.; Neopane, H.P.; Zhu, B. A review on flow and sediment erosion in guide vanes of Francis turbines. *Renew. Sustain. Energy Rev.* **2017**, *75*, 1054–1065. [CrossRef]
9. Qian, Z.; Gao, Y.; Zhang, K.; Huai, W.; Wu, Y. Influence of dynamic seals on silt abrasion of the impeller ring in a centrifugal pump. *Proc. Inst. Mech. Eng. Part A J. Power Energy* **2013**, *227*, 557–566. [CrossRef]
10. Zhang, Y.; Qian, Z.; Ji, B.; Wu, Y. A review of microscopic interactions between cavitation bubbles and particles in silt-laden flow. *Renew. Sustain. Energy Rev.* **2016**, *56*, 303–318. [CrossRef]
11. Smirnov, P.E.; Menter, F.R. Sensitization of the SST turbulence model to rotation and curvature by applying the Spalart–Shur correction term. *J. Turbomach.* **2009**, *131*, 041010. [CrossRef]
12. Guo, B.; Xiao, Y.; Rai, A.K.; Zhang, J.; Liang, Q. Sediment-laden flow and erosion modeling in a Pelton turbine injector. *Renew. Energy* **2020**, *162*, 30–42. [CrossRef]
13. Menter, F.R. Two-equation eddy-viscosity turbulence models for engineering applications. *AIAA J.* **1994**, *32*, 1598–1605. [CrossRef]
14. Xianbei, H.; Qiang, G.; Baoyun, Q. Prediction of Air-Entrained Vortex in Pump Sump: Influence of Turbulence Models and Interface-Tracking Methods. *J. Hydraul. Eng.* **2020**, *146*, 04020010.

15. Celik, I.B.; Ghia, U.; Roache, P.J. Procedure for estimation and reporting of uncertainty due to discretization in CFD applications. *J. Fluids Eng.* **2008**, *130*, 078001.
16. Zeise, B.; Liebich, R.; Prölß, M. Simulation of fretting wear evolution for fatigue endurance limit estimation of assemblies. *Wear* **2014**, *316*, 49–57. [CrossRef]
17. Shrestha, U.; Chen, Z.; Park, S.H.; Do Choi, Y. Numerical studies on sediment erosion due to sediment characteristics in Francis hydro turbine. *IOP Conf. Ser. Earth Environ. Sci.* **2019**, *240*, 042001. [CrossRef]
18. Shen, Z.; Chu, W.; Li, X. Sediment erosion in the impeller of a double-suction centrifugal pump—A case study of the Jingtai Yellow River Irrigation Project, China. *Wear* **2019**, *422–423*, 269–279. [CrossRef]
19. Gautam, S.; Neopane, H.P.; Thapa, B.S.; Chitrakar, S.; Zhu, B. Numerical Investigation of the Effects of Leakage Flow From Guide Vanes of Francis Turbines using Alternative Clearance Gap Method. *J. Appl. Fluid Mech.* **2020**, *13*, 1407–1419.
20. Song, X.J.; Liu, C. Experimental investigation of floor-attached vortex effects on the pressure pulsation at the bottom of the axial flow pump sump. *Renew. Energy* **2020**, *145*, 2327–2336. [CrossRef]
21. Fan, M.; Li, Y.; Ji, P. Energy Characteristics of Full Tubular Pump Device with Different Backflow Clearances Based on Entropy Production. *Appl. Sci.* **2021**, *11*, 3376.
22. Zhao, W.; Egusquiza, M.; Estevez, A. Valentín Improved damage detection in Pelton turbines using optimized condition indicators and data-driven techniques. *Struct. Health Moni.* **2020**, *20*, 3239–3251.
23. Karakas, E.S.; Watanabe, H.; Aureli, M.; Evrensel, C.A. Cavitation Performance of Constant and Variable Pitch Helical Inducers for Centrifugal Pumps: Effect of Inducer Tip Clearance. *J. Fluids Eng.* **2020**, *142*, 1–19. [CrossRef]
24. Cao, J.; Luo, Y.; Presas, A.; Ahn, S.H.; Wang, Z.; Huang, X.; Liu, Y. Influence of rotation on the modal characteristics of a bulb turbine unit rotor. *Renew. Energ.* **2022**, *187*, 887–895. [CrossRef]
25. Takaffoli, M.; Papini, M. Numerical simulation of solid particle impacts on Al6061-T6 Part II: Materials removal mechanisms for impact of multiple angular particles. *Wear* **2012**, *296*, 648–655. [CrossRef]
26. Zhu, Y.; Lu, J.; Liao, H.; Wang, J.; Fan, B.; Yao, S. Research on cohesive sediment erosion by flow: An overview. *Sci. China Ser. E Technol. Sci.* **2008**, *51*, 2001–2012. [CrossRef]

 water

Article

Analysis of Energy Characteristics and Internal Flow Field of "S" Shaped Airfoil Bidirectional Axial Flow Pump

Chuanliu Xie [1,*], Andong Feng [1], Tenglong Fu [1], Cheng Zhang [1], Tao Zhang [1] and Fan Yang [2]

[1] College of Engineering, Anhui Agricultural University, Hefei 230036, China
[2] College of Hydraulic Science and Engineering, Yangzhou University, Yangzhou 214000, China
* Correspondence: xcltg@ahau.edu.cn

Abstract: In order to study the energy characteristics and internal flow field of "S" shaped airfoil bidirectional axial flow pumps, the SST k-ω turbulence model is used to calculate the bidirectional axial flow pump, and the experimental verification is carried out. The results show that the error of numerical calculation of forward and reverse operation is within 5%, and the numerical calculation result is credible. The test results show that the bidirectional axial flow pump has a design flow rate of Q = 368 L/s, head H = 3.767 m, and an efficiency of η = 80.37%. In reverse operation, the flow of the bidirectional axial flow pump under design condition Q = 316 L/s, head H = 3.658 m, efficiency η = 70.37%. The flow of forward operation is about 15% larger than that of reverse operation under design working condition, the design head is about 3.70 m, and the efficiency of design working condition is about 10% higher than that of reverse operation. The numerical calculation results show that under the forward design condition (Q = 368 L/s), the hydraulic loss accounts for 6.22%, and under the reverse design condition (Q = 316 L/s), the hydraulic loss accounts for 11.81%, with a difference of about 6%. The uniformity of impeller inlet flow rate under the forward operation is about 12% higher than that in the reverse operation. In forward and reverse operation, with the increase of flow, the outlet streamline, the outlet total pressure distribution, the uniformity of impeller inlet velocity, and the vortex in the impeller domain are improved, and the forward direction is better than the reverse direction. The research results of this paper can provide a reference for the research and optimal design of the bidirectional axial flow pump.

Keywords: "S" shaped airfoil; bidirectional axial flow pump; energy characteristics; internal flow field

Citation: Xie, C.; Feng, A.; Fu, T.; Zhang, C.; Zhang, T.; Yang, F. Analysis of Energy Characteristics and Internal Flow Field of "S" Shaped Airfoil Bidirectional Axial Flow Pump. *Water* **2022**, *14*, 2839. https://doi.org/10.3390/w14182839

Academic Editors: Changliang Ye, Xijie Song, Ran Tao and Armando Carravetta

Received: 8 August 2022
Accepted: 9 September 2022
Published: 12 September 2022

Publisher's Note: MDPI stays neutral with regard to jurisdictional claims in published maps and institutional affiliations.

Copyright: © 2022 by the authors. Licensee MDPI, Basel, Switzerland. This article is an open access article distributed under the terms and conditions of the Creative Commons Attribution (CC BY) license (https://creativecommons.org/licenses/by/4.0/).

1. Introduction

The water transfer project is a project to improve the people's basic livelihood, playing an increasingly important role in the irrigation of river, coastal, and plain areas. Flood control and drainage standards are also constantly improving, increasing the demand for pumping stations to achieve dual-use irrigation and drainage, and positive and negative performance balance. At present, most of the pumping stations utilize the unidirectional axial flow pump reversing motor operation or switch the river and pumping station flow channel to achieve bidirectional operation. In the former, forward operation of the maximum efficiency of the pump can reach about 85%, and reverse operation because of the vane is a reverse arch state and poor inlet water flow pattern. The flow, head, and efficiency of the pump are greatly weakened, with the maximum efficiency of only 65% or less, showing deviation from national energy-saving and emission reduction targets. Regarding the latter, although the pumping station can realize bidirectional operation and the same performance of forward and reverse operation, the highest efficiency of forward and reverse direction is only about 72%, and the flow channel and river channel are switched. In turn, the civil construction cost is large, which increases the cost of operation and management and maintenance. In response to these problems, the relevant scholars, through the improvement of the axial flow pump and its impeller, support the use of a symmetrical form of wing design bidirectional axial flow pump, in the design method to make innovation. In the development of a new bidirectional axial flow pump impeller, the impeller alone, its forward and reverse performance is superior. At present, the symmetrical wing type has a flat wing type, the center line of the wing type is a straight line, another is "S" shaped airfoil, the center line of the airfoil is "S" shaped. The flat airfoil type bidirectional axial flow pump forward operating efficiency is about 5% worse than the conventional unidirectional axial flow pump forward operating, while reverse operating efficiency is about 15% worse than the conventional unidirectional axial flow pump forward operating, "S" shaped airfoil type bidirectional axial flow pump forward operating efficiency is about 2% worse than the conventional unidirectional axial flow pump forward operating, and reverse operating efficiency is about 10% worse than the conventional unidirectional axial flow pump forward operating. Because of its excellent forward and reverse performance, it has been paid attention to and applied. However, the research on "S" shaped bidirectional axial flow pump is less, and its energy characteristics and internal flow mechanism are not clear, which leads to the lack of theoretical guidance for the design and operation of "S" shaped bidirectional axial flow pumps.

Relevant scholars proposed to use "S" shaped airfoil to design bidirectional axial flow pump [1]. Some scholars have explored the internal flow field of the bidirectional axial flow pump, including the pressure fluctuation characteristics in both forward and reverse directions of the bidirectional axial flow pump [2], the generation mechanism of the vortex in the impeller [3], and the internal flow characteristics of the "S" shaped airfoil bidirectional axial flow pump under cavitation [4], and some scholars have optimized the design of bidirectional axial flow pump or pump device, including reducing the chord length and number of blades and appropriately increasing the axial spacing to enhance the performance of bidirectional axial flow pump [5], investigating the effect of guide vane position on the hydraulic performance and flow pattern of bidirectional vertical shaft cross-flow pump [6], investigating the effect of lobe root clearance on the hydraulic performance of bidirectional axial flow pump [7], and changing the ratio of radius of the centerline circle of S-shaped bend of bidirectional pump to the optimized design of the shaft extension cross-flow pump device of "S" shaped two-way axial flow pump [8]. The influence of different placement positions of the vertical shaft on the performance of the vertical shaft cross-flow pump of "S" shaped two-way axial flow pump was investigated [9]. Some scholars have also studied the flyaway conditions of bidirectional horizontal axial flow pumps operating in forward and reverse directions and concluded that the magnitude of pulsation of axial force in forward operation is significantly larger than that in reverse [10].

The "S" shaped airfoil bidirectional axial flow pump has the advantages of simple structure, convenient installation, easy operation, maintenance, and management, and it has been widely used. In the process of application, relevant scholars have carried out some optimization design work on it or its pump device, but the research on its energy characteristics and internal flow mechanism is less involved, and related studies are also not deep enough. This paper takes the "S" shaped airfoil bidirectional axial flow pump as the research object to explore its energy characteristics and internal flow characteristics during forward and reverse operation, in order to ensure the safe, stable, and efficient operation of the bidirectional axial flow pumping station.

2. Numerical Computation Models, Grids and Computational Methods

2.1. Numerical Calculation Models

The "S" shaped bidirectional axial flow pump is designed by the "S" shaped airfoil, and its geometric structure includes: inlet pipe, impeller, guide vane and outlet pipe, where the number of impeller blades is 4, impeller diameter D is (300–0.2) mm, and the number of guide vane blades is 5. The parameters of the "S" shaped bidirectional axial flow pump are shown in Table 1, the "S" shaped 3D model is shown in Figure 1, and the "S" shaped bidirectional axial flow pump impeller 3D model is shown in Figure 2. The forward and reverse operation of the "S" shaped bidirectional axial flow pump is shown in Figure 3.

Table 1. Parameters of design working conditions of "S" shaped airfoil bidirectional axial flow pump.

Parameters	Forward	Reverse
Blade angle	0°	0°
Impeller diameter	(300–0.2) mm	(300–0.2) mm
Rotating speeds	1450 rpm	1450 rpm
Design flow	368 L/s	316 L/s
Design head	3.70 m	3.70 m
Design point ratio speed	1187	1124

Figure 1. "S" shaped airfoil.

Figure 2. Impeller of bidirectional axial flow pump.

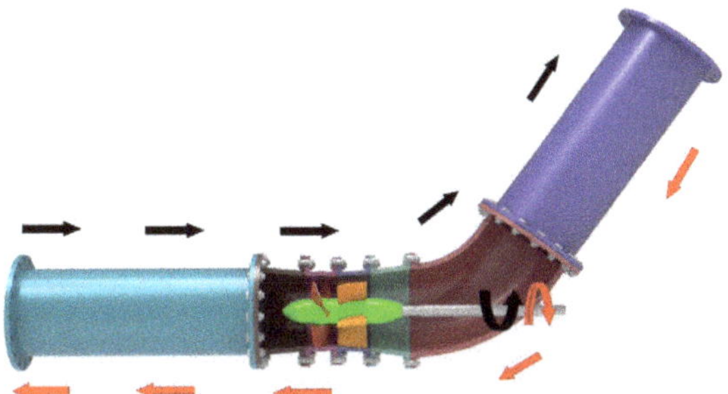

Figure 3. Schematic diagram of forward and reverse operation of "S" shaped airfoil bidirectional axial flow pump. (Black indicates forward operation, Red indicates reverse operation).

As shown in Figure 3, when the bidirectional axial flow pump is in forward operation, the water enters from the side of the inlet pipe and passes through the impeller to lift the energy. Then, the guide vane rectifies the flow, enters the outlet elbow, and then flows out in an oblique 60° direction. In reverse operation, the water enters from the elbow side, flows through the guide vane first, then passes through the impeller to raise energy, and finally flows out of the horizontal straight pipe. In forward and reverse operation, the impeller rotates in the opposite direction.

2.2. Meshing

The fluid inside the inlet and outlet pipes is extracted, the inlet and outlet pipes are divided into unstructured meshes in ANSYS ICEM software, and the boundary layer is locally encrypted considering the thickness of the boundary layer (the change rate of the mesh from the boundary to the interior is 1.05). The impeller and guide vane body are modeled and the structured mesh is divided using ANSYS Turbo Grid software, and the boundary layer and local features of each computational domain are encrypted when the mesh is divided. The impeller adopts "J" topology and guide vane adopts "O" topology. The impeller and guide vane y+ (y+ is a dimensionless quantity of distance from the wall, which is proportional to the height of the first grid layer of the wall. In the numerical calculation using SST k-ω and RNG k-ε turbulence models, the rotational and shear flow y+ is taken as 30~100) values are all around 50, and the impeller top clearance adopts the "H" topology and arranges a 7-layer grid, with y+ values around 10 and grid masses greater than 0.35. The overall grid of the computational model is shown in Figure 4, and the grid irrelevance analysis is shown in Table 2.

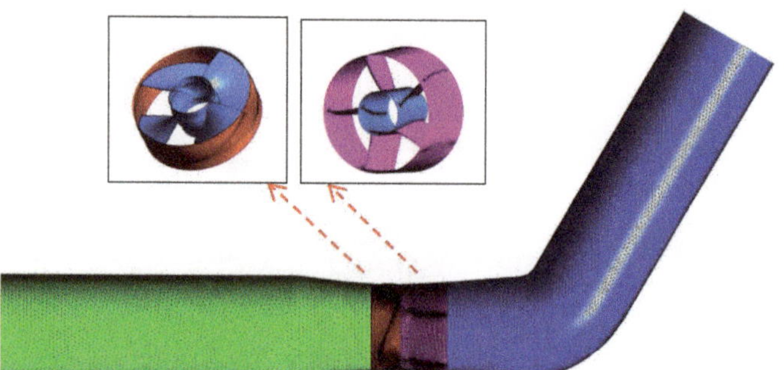

Figure 4. Grid diagram of each component.

Table 2. Grid irrelevance analysis table (Forward running, Q = 400 L/s).

Serial Number	N	η (%)	Serial Number	N	η (%)
1	1,607,288	76.4150	4	3,402,093	77.0300
2	2,143,050	76.4745	5	4,018,220	77.0360
3	2,678,813	77.0405	6	/	/

As shown in Table 2, when the number of grids reaches 2,678,813, the number of grids has basically no effect on the calculation results. Considering the computing power of the computer, grid Serial Number scheme 3 is selected for the subsequent numerical calculation work in this paper.

2.3. Control Equations, Boundary Conditions and Calculation Methods

In this paper, the SST k-ω turbulence model is used to numerically calculate the bidirectional axial flow pump in CFX software. The rated speed of the impeller is set to be −1450 r/min when the bidirectional axial flow pump operates in the forward direction, and 1450 r/min when the impeller operates in the reverse direction. The inlet is set to be the Mass Flow Inlet, the outlet is set to be the Average Static Pressure Outlet, the pressure is set to be 0.2 atm, the Non Moving Wall surface of the solid is set to be the Static Wall Surface, the Non Slip condition is applied, and the boundary condition of the standard wall function is used in the Near Wall Area. The "Stage" interface method is adopted for the dynamic and static interface. The finite volume method based on the finite element is used for discretization of the governing equation, the diffusion term, and pressure gradient are expressed by the finite element function, and the convection term is expressed by the High Resolution Scheme. The convergence condition of each parameter of the flow field is set to 10^{-6}. In principle, the smaller the residual value, the better.

2.4. Analytical Formulae for Numerical Calculation Results

The numerical calculation head H_{net} calculation formula is [11–13]:

$$H_{net} = \left(\frac{\int_{S_2} p_2 u_{t2} dS}{\rho Q g} + H_2 + \frac{\int_{S_2} {u_2}^2 u_{t2} dS}{2Qg}\right) - \left(\frac{\int_{S_1} p_1 u_{t1} dS}{\rho Q g} + H_1 + \frac{\int_{S_1} {u_1}^2 u_{t1} dS}{2Qg}\right) \quad (1)$$

The numerical calculation efficiency η calculation formula is [14,15]:

$$\eta = \frac{\rho g Q H_{net}}{T\omega} \times 100\%, \quad (2)$$

where: P_1 and P_2 are the average static pressure (Pa) at the inlet and outlet of the axial flow pump channel; ρ is the flow density (kg/m^3); g is the acceleration of gravity (m/s^2); s_1 and s_2 are the cross section areas of inlet and outlet of axial flow pump (m^2); u_1 and u_2 are the flow velocity at each point of the inlet and outlet channel section of the axial flow pump (m/s); u_{t1} and u_{t2} are the normal components of flow velocity at each point of the inlet and outlet channel section of axial flow pump (m/s); Q is axial flow pump flow (m^3/s); T is the rotating torque of impeller (N·m); ω is the impeller rotation angle speed (rad/s).

3. Test Device and Test Method

3.1. Test Device

The test bench is a vertical closed cycle system, as shown in Figure 5.

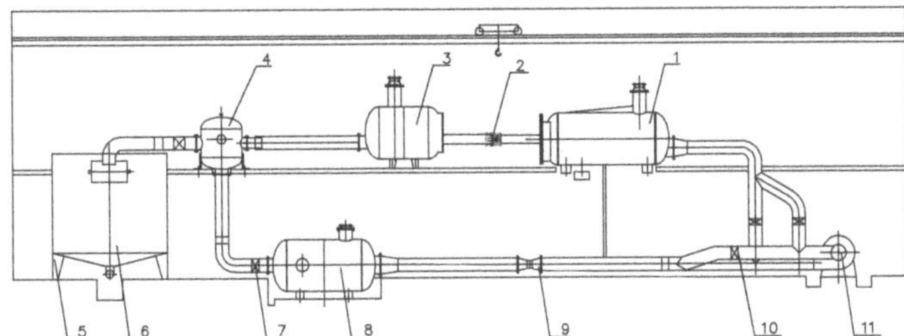

Figure 5. Plan of high precision hydraulic machinery test bench. 1. Water inlet tank; 2. Tested pump; 3. Pressure outlet tank; 4. Bifurcated water tank; 5~6. Flow in situ calibration device; 7. Working condition regulating gate valve; 8. Pressure stabilizing rectifier cylinder; 9. Electromagnetic flowmeter; 10. Operation control gate valve; 11. Auxiliary pump unit.

In the test, the head is measured using the differential pressure transmitter (accuracy $\pm$ 0.015%), the flow is measured by electromagnetic flowmeter (accuracy $\pm$ 0.18%), the speed and torque are measured by speed torque sensor (accuracy $\pm$ 0.24%), the *NPSH* is measured by absolute pressure transmitter (accuracy $\pm$ 0.015%), and the comprehensive uncertainty of the test bench is $\pm$0.39%.

3.2. Test Method

The head H of the test pump is calculated by the following formula [16]:

$$H = \left(\frac{p_2}{\rho g} - \frac{p_1}{\rho g} + z_2 - z_1\right) + \left(\frac{u_2^2}{2g} - \frac{u_1^2}{2g}\right), \tag{3}$$

The test shaft power N is calculated by the following formula [17]:

$$N = \frac{\pi}{30} n(M - M'), \tag{4}$$

Test pump efficiency η calculated by the following formula [18]:

$$\eta = \frac{\rho g Q H}{N} \times 100\%, \tag{5}$$

where H is the pump head (m), P_1 and P_2 are the static pressure (Pa) at the inlet and outlet of the flow field, z_1 and z_2 are the height (m) of the inlet and outlet of the flow field, u_1 and u_2 are the flow velocity (m/s) at the inlet and outlet of the flow field, ρ is the real-time water density of the test (kg/m^3), g is the local gravity acceleration (m/s^2), N is the shaft power (kw), M is the pump input torque (N·m), M' is the pump mechanical loss torque (N·m), n is the pump test speed (r/min), η is the pump model efficiency (%), and Q is the pump flow (m^3/s).

4. Numerical Calculation Results and Experimental Verification

4.1. Experimental Verification of Numerical Calculation Energy Characteristics

The comparison between the forward and reverse numerical calculation and the experimental energy characteristic curve of the "S" shaped airfoil bidirectional axial flow pump is shown in Figure 6.

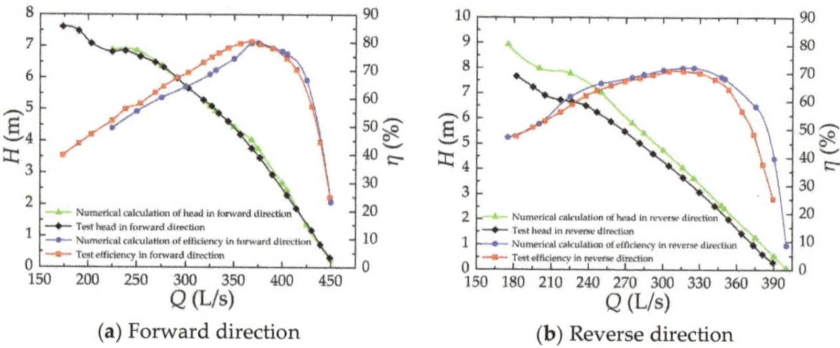

Figure 6. Comparison curve of numerical calculation and test energy characteristics.

By comparing and analyzing the test results in Figure 6, it can be concluded that the $Q = 368$ L/s, $H = 3.767$ m, and efficiency of the bidirectional axial flow pump under the design condition of forward operation $\eta = 80.37\%$, whereas in reverse operation, the flow of bidirectional axial flow pump under design condition $Q = 316$ L/s, $H = 3.658$ m, $\eta = 70.37\%$. The flow of forward operation is about 15% higher than that of reverse operation under the design condition, but the difference in head is small, both of which are about 3.70 m in the design head, and the efficiency of design condition is about 10% higher in the forward direction than in the reverse direction. The main reason why the forward operation performance is better than the reverse operation is that the flow velocity uniformity at the inlet of the forward operation impeller is higher than that in the reverse operation, and the forward water inflow condition is better. The reverse operation has poor water inflow condition due to the front guide vane at the inlet of the impeller.

By comparing and analyzing Figure 6, we can observe that the prediction of forward numerical calculation of bidirectional axial flow pump is more accurate, and the error is basically within 3%, while the prediction accuracy of reverse numerical calculation is slightly worse, and the error is basically within 5%, which means that the numerical calculation accuracy is higher and the calculation results are credible. It can also be found from the figure that the prediction accuracy of the forward numerical calculation of the bidirectional axial flow pump is better than the reverse, which is mainly because the internal flow characteristics of the bidirectional axial flow pump are complicated during reverse operation, so the prediction accuracy is slightly worse.

4.2. Numerical Calculation and Energy Characteristic Analysis

According to the numerical calculation results, the energy characteristic curves of the bidirectional axial flow pump impeller in the forward and reverse directions are organized by Equations (1) and (2), as shown in Figure 7a,b, and the energy characteristics of the bidirectional axial flow pump in the forward and reverse directions are organized as shown in Figure 7c,d.

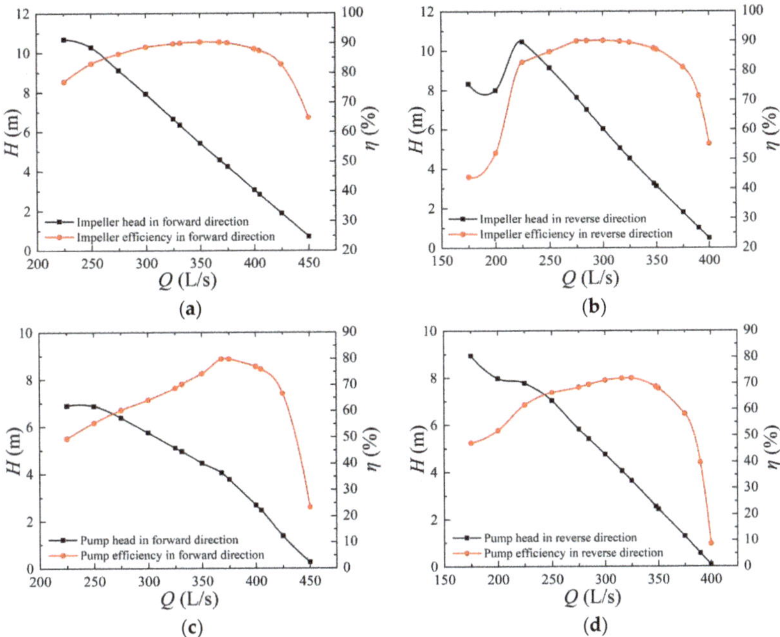

Figure 7. Numerical calculation curve of energy characteristics: (**a**) Forward bidirectional pump impeller; (**b**) Reverse bidirectional pump impeller; (**c**) Forward bidirectional pump; (**d**) Reverse bidirectional pump.

It can be obtained from the numerical calculation that the design working flow rate of the impeller in forward operation is Q = 368 L/s, head H = 4.575 m, efficiency η = 90.32%, and the design working flow rate of the impeller in reverse operation is Q = 316 L/s, head H = 5.055 m, efficiency η = 89.87%. The "S" shaped bidirectional airfoil is completely symmetrical in both forward and reverse directions, and the impeller designed by it is also completely symmetrical in both forward and reverse directions with the same performance, so the difference between the optimal efficiency of the impeller in forward and reverse directions is not large. However, the increase of the guide vane and elbow and other flow guide structure not only makes the "S" shaped airfoil bidirectional axial flow pump optimal operating point shift, and the flow-head to produce differences also larger, but the forward operation of the head under the same flow is significantly greater than the reverse, mainly because the impeller inlet water impulse angle is different, causing the forward operation of the impeller inlet absolute liquid flow angle to be significantly greater than the reverse operation of the impeller inlet absolute liquid flow angle.

Forward operation bidirectional axial flow pump design condition flow rate Q = 368 L/s, head H = 4.049 m, efficiency η = 79.93%, and reverse operation bidirectional axial flow pump design condition flow rate Q = 316 L/s, head H = 4.045 m, efficiency η = 71.91%. By comparing the numerical calculation results, it can be seen that the forward and reverse high flow conditions performance difference is large and forward design condition point efficiency is 8% larger than the reverse. Under the same flow, forward head is equivalent to the reverse vane angle increased by about 2° of head.

There are structural differences between the two directions of operation of the bidirectional axial flow pump. The front of the impeller in forward operation is a straight tube, the impeller outlet is a guide vane and elbow, the reverse operation of the impeller inlet is elbow and guide vane, and the outlet is a straight tube. These structural characteristics of the differences lead to hydraulic losses of the inlet and outlet pipes and guide vane. The flow velocity uniformity and the water impulse angle of the impeller inlet are all different, so the difference between the forward and reverse performance is large.

In this paper, the hydraulic loss ratio is defined as the ratio of the sum of the total hydraulic loss of the inlet pipe, outlet pipe, and guide vane in comparison with the head of the bidirectional axial flow pump, and the hydraulic loss ratio is calculated by the following equation:

$$Ch_f = \frac{h_i + h_g + h_o}{H}, \tag{6}$$

where Ch_f is the hydraulic loss ratio, h_i is the inlet pipe hydraulic loss (m); h_g is the guide vane body hydraulic loss (m); h_o is the outlet pipe hydraulic loss (m); H is the pump head (m).

The equation for the inlet flow velocity distribution uniformity of the impeller is [19]:

$$V_u = \left\{1 - \frac{1}{\overline{v_a}}\sqrt{\left[\sum_{i=1}^{n}(v_{ai} - \overline{v_a})^2\right]/n}\right\} \times 100\%, \tag{7}$$

where V_u is the uniformity of axial flow velocity distribution in the characteristic section (%); v_{ai} is the axial velocity of each calculation unit (m/s); n is the number of calculation units. v_a is the axial flow velocity at impeller inlet.

Forward and reverse operation bidirectional axial flow pump inlet and outlet pipes and guide vane hydraulic loss are shown in Figure 8. The impeller inlet flow rate uniformity is shown in Figure 9.

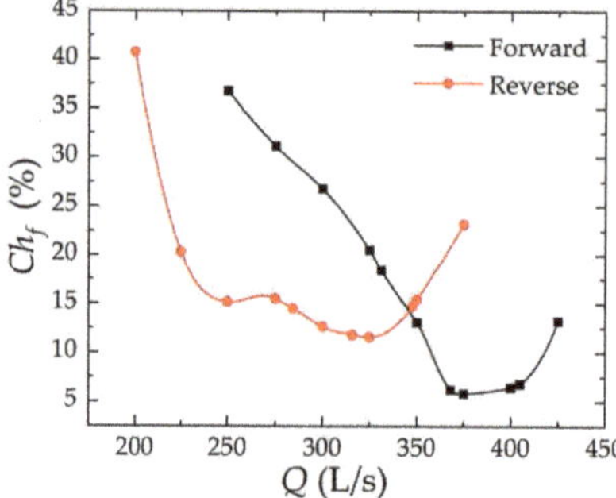

Figure 8. Proportion of hydraulic loss.

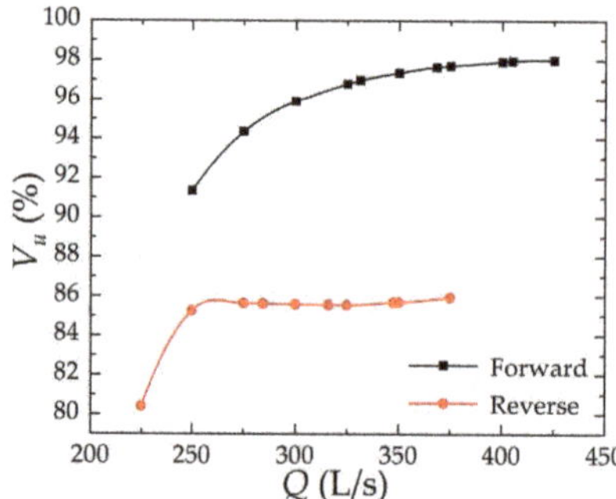

Figure 9. Uniformity of flow velocity at impeller inlet.

Through Figures 8 and 9, it can be concluded that the, considering forward design condition (Q = 368 L/s), the hydraulic loss percentage in the peak and valley region, the value is close to the minimum 6.22%, and considering the reverse design condition

(Q = 316 L/s), the hydraulic loss percentage is also in the peak and valley region. The value is close to the minimum 11.81%, with a difference of about 6%, and this leads to the reverse design condition efficiency being lower than the forward 6%. The efficiency of the forward design working point is 8% larger than that of the reverse, and the other 2% difference is mainly due to the fact that the inlet flow velocity uniformity of the impeller in the forward operation is about 12% higher than that of the reverse, resulting in a difference of about 2% in the forward and reverse efficiency. The main reason for the different hydraulic loss and flow velocity uniformity in forward and reverse operation is that in reverse operation, the curved guide vane is front-loaded, which reduces the flow velocity uniformity of the impeller inlet, resulting in an increase in the bad flow pattern of the impeller inlet and outlet and an increase in the proportion of hydraulic loss.

4.3. Analysis of Internal Flow Fields for Numerical Calculations

The overall streamline of the bidirectional axial flow pump forward taking $0.9Q_{df}$, $1.0Q_{df}$, $1.1Q_{df}$ (Forward design working flow rate Q_{df} = 368 L/s) is shown in Figures 10a, 11a and 12a, and the overall streamline of the bidirectional axial flow pump reverse taking $0.9Q_{dr}$, $1.0Q_{dr}$, $1.1Q_{dr}$ (Reverse design working flow rate of Q_{dr} = 316 L/s) is shown in Figures 10b, 11b and 12b. As shown in Figures 10–12, forward operation of the inlet pipe for the straight pipe and reverse operation of the inlet pipe for the elbow, forward, and reverse operation of the inlet water flow state are better and the streamline is uniform.

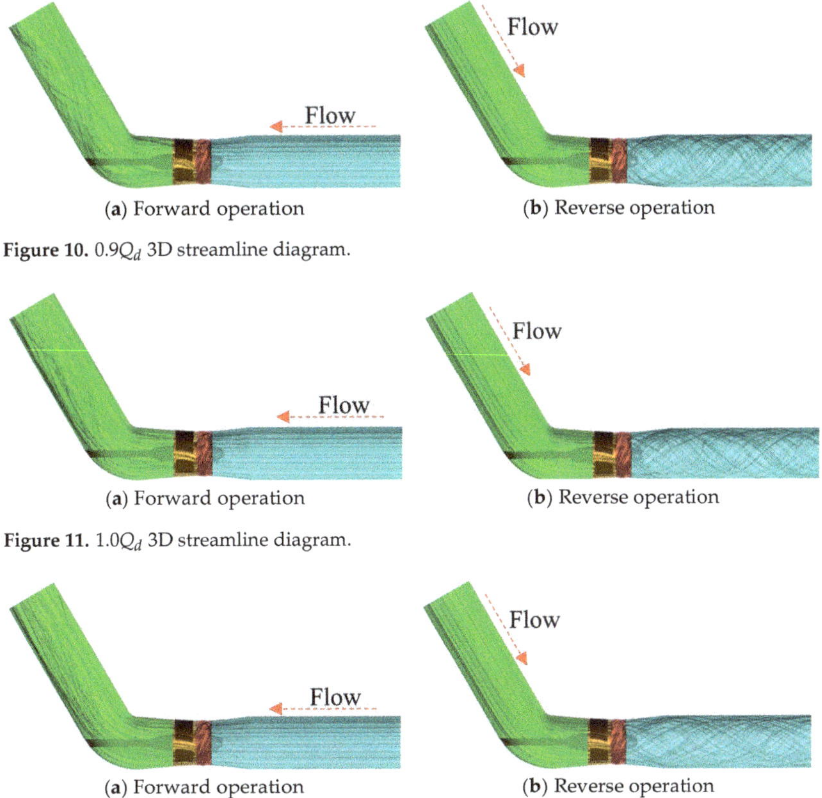

(**a**) Forward operation (**b**) Reverse operation

Figure 10. $0.9Q_d$ 3D streamline diagram.

(**a**) Forward operation (**b**) Reverse operation

Figure 11. $1.0Q_d$ 3D streamline diagram.

(**a**) Forward operation (**b**) Reverse operation

Figure 12. $1.1Q_d$ 3D streamline diagram.

In forward operation, considering the outlet elbow with the increase in flow, the phenomenon of cross-winding of the streamline in the outlet elbow is gradually weakened, whereas in reverse operation, because the impeller releases water directly into the outlet straight pipe, there is no guide leaf recovery ring volume, resulting in the water out of the water flow rotating out, leading to increased hydraulic losses or an increase in the circumferential velocity of the water flow, resulting in kinetic energy affecting the ability to reduce the pressure energy, so the head in reverse operation is lower than the head of forward operation. The impeller inlet is affected by the inlet elbow and guide vane under reverse operation conditions, resulting in low flow velocity uniformity of the impeller inlet and increasing the hydraulic loss of the inlet structure. Forward operation inlet water state is slightly better than reverse operation. Outflow water state forward operation is significantly greater than reverse operation, which leads to reverse operation total hydraulic loss being significantly greater than forward operation. Forward and reverse operation energy performance curves are compared to the same flow conditions. Forward operation head and efficiency is significantly higher than reverse operation (shown in Figure 6).

The total pressure distribution clouds of the middle section of the fluid calculation domain for both forward and reverse directions of the bidirectional axial flow pump taking $0.9Q_d$, $1.0Q_d$, $1.1Q_d$ are shown in Figures 13–15.

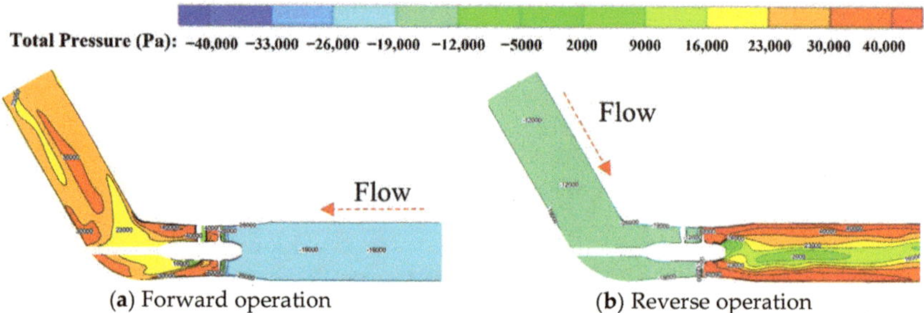

(**a**) Forward operation (**b**) Reverse operation

Figure 13. $0.9Q_d$ clouds of total pressure distribution in the middle cross section.

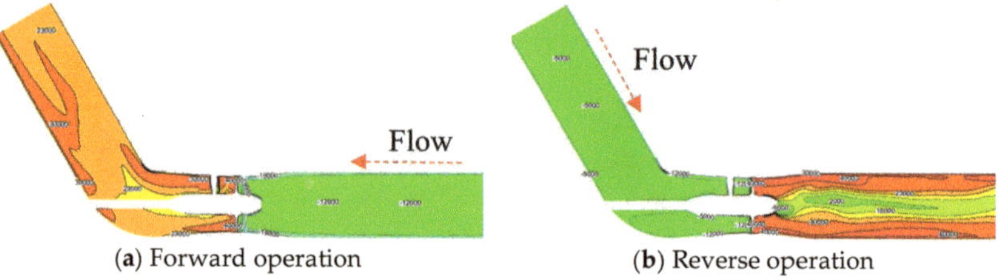

(**a**) Forward operation (**b**) Reverse operation

Figure 14. $1.0Q_d$ clouds of total pressure distribution in the middle cross section.

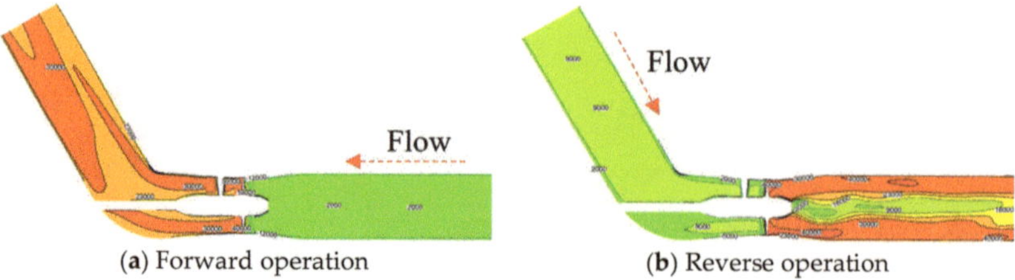

(**a**) Forward operation (**b**) Reverse operation

Figure 15. $1.1Q_d$ clouds of total pressure distribution in the middle cross section.

According to Figures 13–15, the total pressure distribution in the inlet pipe is relatively uniform under forward and reverse operation. During forward operation, the total pressure distribution in the outlet elbow becomes more and more uniform with the increase of flow.

During reverse operation, due to the rotation of the impeller, there is no circulation recovery structure such as guide vane at the impeller outlet, resulting in uneven total pressure distribution in the center of the outlet, obvious stratification, low total pressure in the middle area, and high total pressure at the side wall of the outlet pipe. With the increase of flow rate, the distribution uniformity of total pressure of outlet water is improved. Forward operation and reverse operation affect the size of the hydraulic loss of the inlet pipe and the recovery of more and less water pressure energy from the outlet pipe. Considering the head under the same flow condition, through the static pressure distribution cloud diagram, it can be concluded that the forward running inlet pipe static pressure distribution is more uniform than the reverse running, so the forward running inlet pipe hydraulic loss is less than the reverse running. The hydrostatic pressure distortion area in the forward running outlet bend is less than the reverse running, and the outlet pressure recovery is more than the reverse running, as reflected in the energy performance curve (shown in Figure 7c,d). At the same flow rate, the head of forward operation is higher than that of reverse operation.

The axial flow velocity distribution clouds of the impeller inlet for both forward and reverse directions of the bidirectional axial flow pump are taken as $0.9Q_d$, $1.0Q_d$, and $1.1Q_d$ as shown in Figures 16–18.

As can be obtained from Figures 16–18, the uniformity of the axial flow velocity of the impeller inlet increases with the increase of the flow rate under both forward and reverse operation. In forward operation, because the number of impeller blades is 4, the impeller inlet axial velocity is obviously divided into 4 areas, and in reverse operation, the impeller inlet is not only influenced by the number of impeller blades, but also by the number of guide vanes, and the impeller inlet axial velocity is divided into 4 areas near the hub and 5 areas near the rim. Overall, the uniformity of axial velocity at the inlet of forward running impeller is higher than that of the reverse running impeller. This conclusion can also be explained by Figure 9.

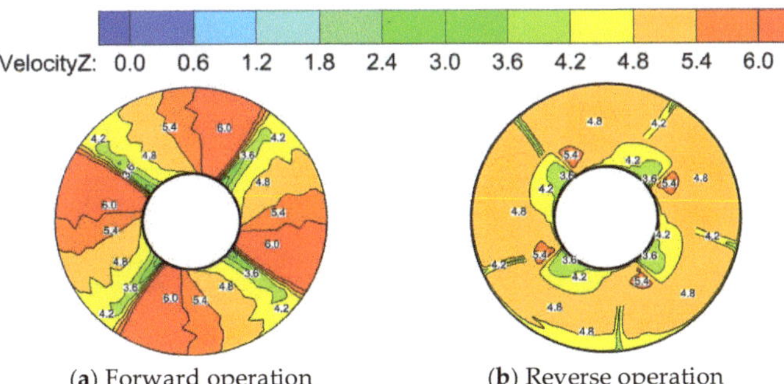

(**a**) Forward operation (**b**) Reverse operation

Figure 16. $0.9Q_d$ impeller inlet axial flow velocity distribution cloud diagram.

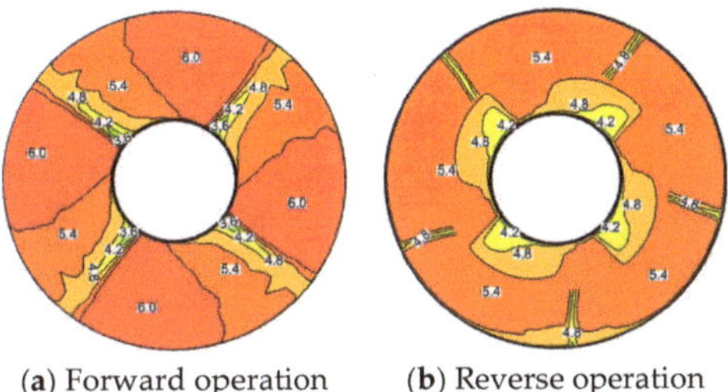

(**a**) Forward operation (**b**) Reverse operation

Figure 17. $1.0Q_d$ impeller inlet axial flow velocity distribution cloud diagram.

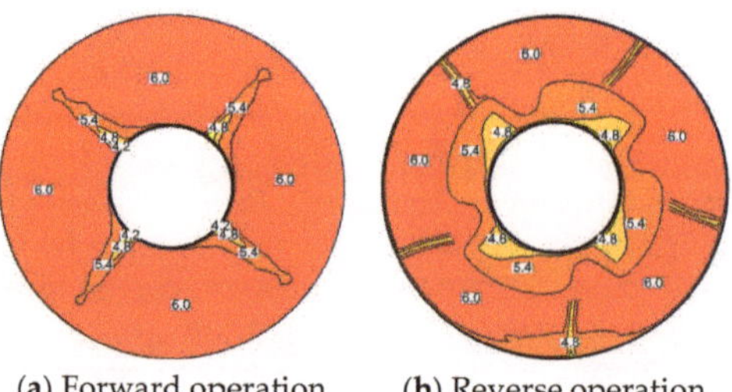

(a) Forward operation (b) Reverse operation

Figure 18. $1.1Q_d$ impeller inlet axial flow velocity distribution cloud diagram.

The vortex distribution inside the impeller for the forward and reverse directions of the bidirectional axial flow pump taking $0.9Q_d$, $1.0Q_d$, and $1.1Q_d$ is shown in Figures 19–21. In this paper, the vortex discrimination criterion is Q-Criterion, which was proposed by Hunt et al. in 1988 [20,21].

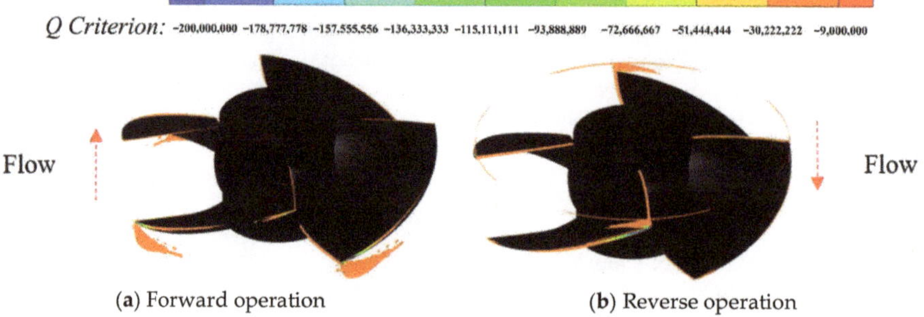

(a) Forward operation (b) Reverse operation

Figure 19. $0.9Q_d$ impeller internal vorticity distribution diagram.

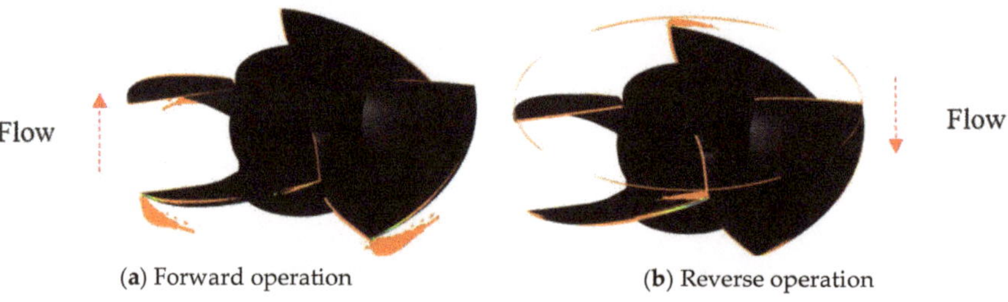

(a) Forward operation (b) Reverse operation

Figure 20. $1.0Q_d$ impeller internal vorticity distribution diagram.

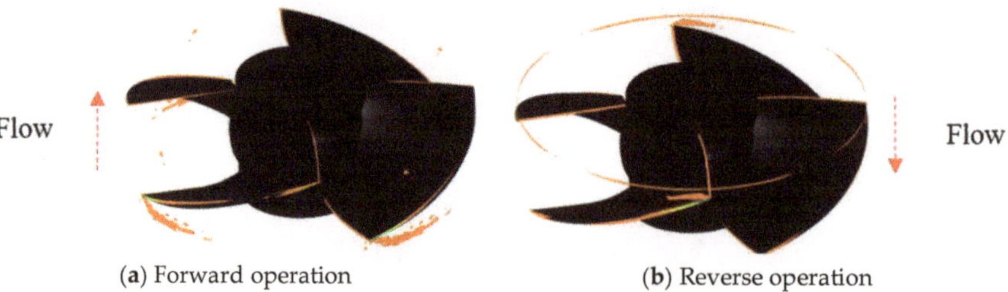

(a) Forward operation (b) Reverse operation

Figure 21. $1.1Q_d$ impeller internal vorticity distribution diagram.

As shown in Figures 19–21, forward operation compared to reverse, impeller blade forward, and reverse pressure difference is larger. So, in the blade of the suction surface slightly longer leakage vortex is indicates, whereas in the corresponding reverse operating conditions, the blade of slightly pointed leakage vortex is relatively slightly small. The area of tip clearance vortex in reverse operation is slightly larger than that in forward operation. Due to the front guide vane and elbow in reverse operation, the vortex area at the inlet side of the blade is also significantly larger than that in forward operation. At the same time, the reverse operation also causes an annular vortex belt near the rim between the guide vane and the impeller. In both forward and reverse operation, the vortex area decreases with the increase of flow. This also reflects the higher energy performance of the forward operating impeller compared to the reverse (shown in Figure 7a,b).

5. Conclusions

This paper reveals the forward and reverse energy and internal flow characteristics of the bidirectional axial flow pump through numerical calculations and experimental tests on the bidirectional axial flow pump, and the following conclusions are obtained:

(1) The comparative analysis of the numerical calculation results and tests of the energy characteristics of the bidirectional axial flow pump shows that the predictions of the forward and reverse numerical calculations are relatively accurate, and the error is basically within 5%. Compared with the forward prediction, the accuracy of the reverse numerical calculation is slightly worse, and the numerical calculation results are credible.

(2) The test results show that the bidirectional axial flow pump design working condition flow rate $Q = 368$ L/s, head $H = 3.767$ m, and efficiency $\eta = 80.37\%$ in forward operation and bidirectional axial flow pump design working condition flow rate $Q = 316$ L/s, head $H = 3.658$ m, efficiency $\eta = 70.37\%$ in reverse operation. The forward operation is about 15% larger than the reverse operation design working condition flow rate, the design head is about 3.70 m, and the design working efficiency is about 10% higher in the forward direction than in the reverse direction.

(3) The numerical calculation results show that under the forward design condition ($Q = 368$ L/s), the proportion of hydraulic loss is 6.22%, and under the reverse design condition ($Q = 316$ L/s), the proportion of hydraulic loss is 11.81%, with a difference of about 6%. The uniformity of impeller inlet flow velocity is about 12% higher than that under the reverse operation. The main reason for the difference in hydraulic loss and flow velocity uniformity between forward and reverse operation is that during reverse operation, curved guide vanes are placed in front, which reduces the flow velocity uniformity at the inlet of the impeller, resulting in an increase in the bad flow pattern of the inlet water of the impeller, and because the outlet water has no circulation recovery structure such as guide vanes, the ability of converting kinetic energy into pressure energy of the outlet water is weakened, and the proportion of hydraulic loss is increased.

(4) In the forward operation, the inlet water is straight pipe, and in the reverse operation, the inlet water is elbow. Under the forward and reverse operation, the inlet water flow pattern is relatively good. In the forward and reverse operation, with the increase of flow, the outlet water streamline, the total pressure distribution of outlet water, the uniformity of impeller inlet flow velocity, and the vortex in the impeller domain are improved. The internal flow fields, such as outlet streamline, total outlet pressure distribution, impeller inlet velocity uniformity, and impeller domain vortex, under forward operation are better than those under reverse operation, so the performance of forward operation is better than that of reverse operation.

Author Contributions: Concept design, C.X. and A.F.; numerical calculation, T.F. and C.Z.; experiment and data analysis, T.Z. and F.Y.; manuscript writing, C.X. and A.F. All authors have read and agreed to the published version of the manuscript.

Funding: This research was funded by Anhui Province Natural Science Funds for Youth Fund Project, grant number 2108085QE220. Key scientific research project of Universities in Anhui Province, grant number KJ2020A0103. Anhui Province Postdoctoral Researchers' Funding for Scientific Research Activities, grant number 2021B552. Anhui Agricultural University President's Fund, grant number 2019zd10. Stabilization and Introduction of Talents in Anhui Agricultural University Research Grant Program, grant number rc412008.

Data Availability Statement: Not applicable.

Conflicts of Interest: The authors declare no conflict of interest.

References

1. Tang, F.; Liu, C.; Wang, G.; Xie, W.; Zhou, J.; Cheng, L. Study on hydraulic model of bidirectional axial flow pump with plane S-shaped channel. *J. Agric. Mach.* **2003**, *34*, 50–53.
2. Ma, P.; Wang, J.; Li, H. Numerical Analysis of Pressure Pulsation for a Bidirectional Pump under Positive and Reverse Operation. *Adv. Mech. Eng.* **2014**, *6*, 730280. [CrossRef]
3. Ma, P.; Wang, J. An analysis on the flow characteristics of bi-directional axial-flow pump under reverse operation. *Proc. Inst. Mech. Eng. Part A J. Power Energy* **2017**, *231*, 239–249. [CrossRef]
4. Xie, C. Study on Energy and Cavitation Characteristics of S-Shaped Airfoil Bidirectional Axial Flow Pump. Ph.D. Thesis, Yangzhou University, Yangzhou, China, 2018.
5. Ma, P.; Wang, J. Influence of geometric parameters of straight guide vane on flow field and hydraulic performance of bidirectional pump. *J. Hydraul. Eng.* **2017**, *48*, 1126–1133.
6. Meng, F.; Pei, J.; Li, Y.; Yuan, S.; Chen, J. Influence of guide vane position on hydraulic performance of bidirectional Shaft Tubular pump device. *J. Agric. Mach.* **2017**, *48*, 135–140.
7. Meng, F.; Li, Y.; Yuan, S.; Yuan, J.; Zheng, Y.; Yang, P. Effect of blade root clearance on hydraulic performance of bidirectional axial flow pump. *J. Agric. Mach.* **2020**, *51*, 131–138.
8. Zheng, Y.; Li, C.; Gu, X.; Chen, Y. Effect of S-shaped elbow on performance and stability of bidirectional axial extension pump. *J. Eng. Thermophys.* **2019**, *40*, 319–327.
9. Jin, K.; Chen, Y.; Tang, F.; Shi, L.; Liu, H.; Zhang, W. Influence of shaft position on hydraulic characteristics of two-way tubular pump device. *J. Hydroelectr. Eng.* **2021**, *40*, 67–77.
10. Kan, K.; Zhang, Q.; Xu, Z.; Chen, H.; Zheng, Y.; Zhou, D.; Maxima, B. Study on a horizontal axial flow pump during runaway process with bidirectional operating conditions. *Sci. Rep.* **2021**, *11*, 21834. [CrossRef] [PubMed]
11. Yang, F.; Zhang, Y.; Yuan, Y.; Liu, C.; Li, Z. Numerical and Experimental Analysis of Flow and Pulsation in Hump Section of Siphon Outlet Conduit of Axial Flow Pump Device. *Appl. Sci.* **2021**, *11*, 4941. [CrossRef]
12. Fang, X.; Hou, Y.; Cai, Y.; Chen, L.; Lai, T.; Chen, S. Study on a high-speed oil-free pump with fluid hydrodynamic lubrication. *Adv. Mech. Eng.* **2020**, *12*, 1687814020945463. [CrossRef]
13. Stuparu, A.; Baya, A.; Bosioc, A.; Anton, L.; Mos, D. Experimental investigation of a pumping station from CET power plant Timisoara. In *IOP Conference Series: Earth and Environmental Science*; IOP Publishing: Bristol, UK, 2019; Volume 240.
14. Xie, C.; Tang, F.; Yang, F.; Zhang, W.; Zhou, J.; Liu, H. Numerical simulation optimization of axial flow pump device for elbow inlet channel. In *IOP Conference Series: Earth and Environmental Science*; IOP Publishing: Bristol, UK, 2019; Volume 240.
15. Shi, L.; Zhang, W.; Jiao, H.; Tang, F.; Wang, L.; Sun, D.; Shi, W. Numerical simulation and experimental study on the comparison of the hydraulic characteristics of an axial-flow pump and a full tubular pump. *Renew. Energy* **2020**, *153*, 1455–1464. [CrossRef]
16. Shi, L.; Tang, F.; Zhou, H.; Tu, L.; Xie, R. Axial-flow pump hydraulic analysis and experiment under different swept-angles of guide vane. *Trans. Chin. Soc. Agric. Eng.* **2015**, *31*, 90–95.
17. Zhang, W.; Tang, F.; Shi, L.; Hu, Q.; Zhou, Y. Effects of an Inlet Vortex on the Performance of an Axial-Flow Pump. *Energies* **2020**, *13*, 2854. [CrossRef]
18. Yang, F.; Zhao, H.; Liu, C. Improvement of the Efficiency of the Axial-Flow Pump at Part Loads due to Installing Outlet Guide Vanes Mechanism. *Math. Probl. Eng.* **2016**, *2016*, 6375314. [CrossRef]
19. Sun, Z.; Yu, J.; Tang, F. The Influence of Bulb Position on Hydraulic Performance of Submersible Tubular Pump Device. *J. Mar. Sci. Eng.* **2021**, *9*, 831. [CrossRef]
20. Wang, Y.; Zhang, W.; Cao, X.; Yang, H. The applicability of vortex identification methods for complex vortex structures in axial turbine rotor passages. *J. Hydrodyn.* **2019**, *31*, 700–707. [CrossRef]
21. Tang, X.; Jiang, W.; Li, Q.; Gaoyang, H.; Zhang, N.; Wang, Y.; Chen, D. Analysis of hydraulic loss of the centrifugal pump as turbine based on internal flow feature and entropy generation theory. *Sustain. Energy Technol. Assess.* **2022**, *52*, 102070.

 water

MDPI

Article

Analysis of the Flow Energy Loss and *Q-H* Stability in Reversible Pump Turbine as Pump with Different Guide Vane Opening Angles

Wei Yan [1], Di Zhu [2], Ran Tao [3,4,5] and Zhengwei Wang [5,6,*]

1 ANDRITZ (China) Ltd., Beijing 100004, China
2 College of Engineering, China Agricultural University, Beijing 100083, China
3 College of Water Resources and Civil Engineering, China Agricultural University, Beijing 100083, China
4 Beijing Engineering Research Center of Safety and Energy Saving Technology for Water Supply Network System, China Agricultural University, Beijing 100083, China
5 State Key Laboratory of Hydroscience and Engineering, Tsinghua University, Beijing 100084, China
6 Department of Energy and Power Engineering, Tsinghua University, Beijing 100084, China
* Correspondence: wzw@mail.tsinghua.edu.cn

Abstract: The internal flow problem of a reversible pump turbine restricts its safe and stable operation. Among them, the influence of the guide vane on the internal flow field is very crucial. The flow–head relationship is of great significance in the performance stability of the unit. In this study, the performance and flow field characteristics under different flow rates were analyzed for different guide vane opening angles. By comparing the results of the model test and computational fluid dynamics simulation, it was found that the simulation can well predict the energy characteristics and flow field distribution. There is an optimal efficiency range under each guide vane opening angle. The increase or decrease in flow will reduce the efficiency. For the head, it will decrease significantly with a decrease in the flow rate, especially when it deviates seriously from the optimal efficiency region. From the contour of the flow energy loss and the vector of velocity, it can be seen that the head drop is closely related to the flow blockage caused by the difference between the runner incoming flow direction and the installation direction of the guide vane. This study deeply revealed the valley and peak of head variation under different guide vane opening conditions. It can provide technical support for improving the wide range operation stability of a pump turbine.

Keywords: pump turbine; flow energy loss; flow–head stability; guide vane opening

Citation: Yan, W.; Zhu, D.; Tao, R.; Wang, Z. Analysis of the Flow Energy Loss and *Q-H* Stability in Reversible Pump Turbine as Pump with Different Guide Vane Opening Angles. *Water* **2022**, *14*, 2526. https://doi.org/10.3390/w14162526

Academic Editor: Giuseppe Pezzinga

Received: 5 July 2022
Accepted: 11 August 2022
Published: 17 August 2022

Publisher's Note: MDPI stays neutral with regard to jurisdictional claims in published maps and institutional affiliations.

Copyright: © 2022 by the authors. Licensee MDPI, Basel, Switzerland. This article is an open access article distributed under the terms and conditions of the Creative Commons Attribution (CC BY) license (https://creativecommons.org/licenses/by/4.0/).

1. Introduction

In order to achieve sustainable development, China has proposed the goal of carbon peaking and carbon neutralization, built a modern energy system and enhanced the stability, security and sustainability of the energy supply [1,2]. Pumped storage technology is the most mature, economical and large-scale energy storage technique [3]. It has the functions of peak-load regulation, valley-load filling and phase regulation [4]. It is an important part of realizing the flexible regulation of the power system. A reversible pump turbine is the key component of a pumped storage power station [5]. According to the regulation requirements of the power station, a pump turbine usually switches frequently under multiple working conditions [6]. Therefore, the weighted average flow rate of a pump turbine under multiple working conditions is very important. It determines the speed of the pumping process or power generation process and represents the energy charging and discharging performance of the pumped storage power station.

During the operation of a pump turbine, the weighted average flow rate is strongly related to the opening angle of the guide vane under a certain head, and the opening of the guide vane is greatly affected by the matching of the rotor and stator [7]. For a specific guide vane opening angle, its adaptive optimal flow rate is generally within a specific range, and exceeding this range will cause performance degradation [8]. The flow–head instability problem of a pump turbine with a small flow and high head under a pump condition matches, to some extent, the problem of a runner incoming flow and guide vane opening angle [9,10]. Many scholars have carried out extensive and in-depth research on the flow–head instability and dynamic and static interference of pump turbines. Li et al. [11,12] studied the flow–head instability characteristics of pump turbines by combining an experiment and numerical simulation. The results show that, when entering the unstable area, several vortices are distributed in the guide vane channel and evenly distributed along the circumferential direction. The intensity and range of the vortex change with the flow rate, and the vortex increases the hydraulic loss. Xiao et al. [13] found that the formation of the flow–head unstable area is related to the huge hydraulic loss inside the draft tube, runner and guide vane. There are secondary flow, reflux and even vortex in the unstable area. Lu et al. [14] used a combination of a numerical simulation and experiment to compare the influence of the guide vane profile on the internal flow field of the pump turbine, and found that the guide vane profile has a significant influence on the flow field characteristics, especially the flow energy loss of adjacent components. Song et al. [15] studied the matching between the guide vane opening and the runner and found that, with an increase in the guide vane opening, the magnitude and pulsation amplitude of the runner radial force also increased. Through a pressure pulsation test of the pump turbine, Zhang et al. [16] found that, under a partial load, the rotor–stator interaction between the runner and guide vane will cause the pressure pulsation in the vaneless area.

In the flow–head instability area, the internal flow of the pump turbine is complex [17]. Improving the head-drop margin and increasing the distance between the unstable point and the high efficiency area can improve the operation stability and safety of the unit under pump conditions [18,19]. However, in some cases with a large guide vane opening and low head, if the head rises and the guide vane is not adjusted to the appropriate angle, a flow–head instability phenomenon will also occur [20]. Overcoming this phenomenon, can be realized by adjusting the opening angle of the guide vane, but this method will affect the weighted average flow rate of the pump turbine unit and the economy of the power station operation. In fact, CFD technology and the analysis of the second law of thermodynamics can be used in engineering to help to clarify the change law and reason of flow head characteristics. Many scholars have carried out corresponding research on the correlation analysis of turbomachinery. Zhao et al. [21] described the design and analysis method of a multistage vaneless reverse rotating turbine. The entropy increase in tip leakage is used to describe the cause and result of turbine loss, so as to investigate the performance of the turbine under off design conditions. Based on the concept of entropy, Vanzante et al. [22] introduced the concept of loss work to analyze the loss of the compressor. Sun et al. [23] conducted an unsteady three-dimensional numerical simulation of a transonic compressor. The compressor stage loss and its concentration area were obtained by using the loss work research method. Other scholars use the entropy production rate [24] to measure the energy dissipation in flow. Kluxen et al. [25] analyzed the local entropy production rate in an axial flow turbine and determined that the main factor of additional loss was the strength of the reflux zone. Soltanmohamadi et al. [26] used the entropy production analysis method to optimize the design of the turbine, reducing the energy loss by 26.02% in the full operating range. Zeinalpour et al. [27] also proposed an optimization method for turbine blades based on the entropy yield theory. Taking the entropy production rate as the quantitative index of loss, the optimal design of a multistage turbine cascade was successfully carried out. The above concepts and research methods have great reference value and enlightening significance for the quantitative analysis of the internal flow and energy loss of a pump turbine under pump conditions.

In order to clarify the unit performance degradation under different guide vane opening angles, this study analyzed the flow–head (Q-H) characteristics from a 10 degrees opening and 18 degrees opening for a pump turbine model unit in pump mode. The analysis of the flow field provides a basis for the mechanism of performance degradation, which can clarify the matching problem of a unit runner guide vane under different guide vane openings and provide technical support for pumped storage power stations. The innovation of this paper is to clarify the *Q-H* characteristics in depth by combining the experimental and simulation results. The comparative analysis of various working conditions is relatively rare in practical projects. This paper creatively analyzes the *Q-H* characteristics comprehensively, which is of great help to the stable operation of the unit in a wide range.

2. Research Objective

The research objective is to experimentally investigate a model scale of a pump turbine. It includes the volute, stay vanes, guide vanes, runner and draft tube as shown in Figure 1. The main parameters are shown in Table 1. The blade number of the runner is $Z_r = 9$. Both the guide vane blade number Z_g and stay vane blade number Z_s are 22. The runner diameter (diameter at high pressure side) D_1 is 554 mm. The design head H_d is 52.9 m, the maximum head H_{max} is 58.0 m and the minimum head H_{min} is 50.8 m. The rated rotational speed n_d is 1100 rpm. The specific speed n_q can be calculated by:

$$n_q = \frac{n_d \sqrt{Q_d}}{H_d^{3/4}} \tag{1}$$

In this study, n_q is approximately 34.6.

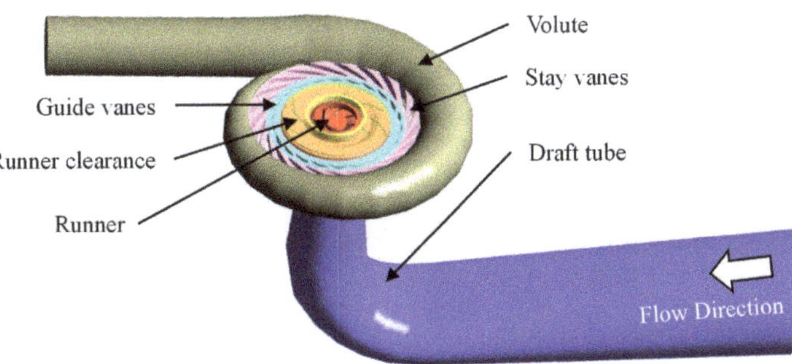

Figure 1. The 3D model of the fluid domain of the reversible pump turbine's hydraulic components.

Table 1. Main parameter of reversible pump turbine unit.

Parameter	Symbol	Value
Runner blade number	Z_r	9
Guide vane blade number	Z_g	22
Stay vane blade number	Z_s	22
Runner diameter	D_1	554 [mm]
Design head	H_d	52.9 [m]
Maximum head	H_{max}	58.0 [m]
Minimum head	H_{min}	50.8 [m]
Rated rotational speed	n_d	1100 [rpm]
Specific speed	n_q	34.6

3. Methodology

3.1. Governing Equations

In this study, computational fluid dynamics (CFD) was used to simulate the internal flow in pump turbine. The flow in the pump turbine can be regarded as three-dimensional incompressible turbulent flow, which is based on Reynolds-averaged Navier–Stokes (RANS) equation and total energy equation. The continuity equation, momentum equation and total energy equation are expressed as follows:

$$\frac{\partial \overline{u_i}}{\partial x_i} = 0 \tag{2}$$

$$\rho \frac{\partial \overline{u_i}}{\partial t} + \rho \overline{u_j} \frac{\partial \overline{u_i}}{\partial x_j} = \frac{\partial}{\partial x_j}\left(-\overline{p}\delta_{ij} + 2\mu \overline{S_{ij}} - \rho \overline{u_i' u_j'}\right) \tag{3}$$

$$\frac{\partial}{\partial t}(\rho h_{tot}) - \frac{\partial p}{\partial t} + \frac{\partial}{\partial x_j}(\rho u_j h_{tot}) = \frac{\partial}{\partial x_j}\left(\lambda_t \frac{\partial T}{\partial x_j} - \overline{u_j h_{sta}}\right) + \frac{\partial}{\partial x_j}\left[u_j\left(2\mu \overline{S_{ij}} - \rho \overline{u_i' u_j'}\right)\right] \tag{4}$$

where u—velocity, t—time, ρ—density, T—temperature, x—coordinate component, δ_{ij}—Kroneker delta, μ—dynamic viscosity, S_{ij}—mean rate of strain tensor, h_{sta}—static enthalpy, h_{tot}—total enthalpy, λ_t—thermal conductivity.

SST k-ω model was used as the turbulence model [28] to close the governing equation. The SST k-ω model is a regional mixed model that is suitable for the simulation of different flows in engineering. It is applicable to strong pressure gradient and strong shear near the wall. The flow in the pump turbine in this paper can be well simulated by saving calculation resources while ensuring calculation accuracy. Turbulent kinetic energy k equation and specific dissipation rate ω equation can be written as:

$$\frac{\partial(\rho k)}{\partial t} + \frac{\partial(\rho u_i k)}{\partial x_i} = P - \frac{\rho k^{3/2}}{l_{k\text{-}\omega}} + \frac{\partial}{\partial x_i}\left[(\mu + \sigma_k \mu_i)\frac{\partial k}{\partial x_i}\right] \tag{5}$$

$$\frac{\partial(\rho \omega)}{\partial t} + \frac{\partial(\rho u_i \omega)}{\partial x_i} = C_\omega P - \beta \rho \omega^2 + \frac{\partial}{\partial x_i}\left[(\mu + \sigma_\alpha \mu_i)\frac{\partial \omega}{\partial x_i}\right] + 2(1 - F_1)\frac{\rho \sigma_{\omega 2}}{\omega}\frac{\partial k}{\partial x_i}\frac{\partial \omega}{\partial x_i} \tag{6}$$

where P—the production term in k and ω equations. F_1—the blending function. σ_k, σ_ω—constants of turbulence model. $l_{k\text{-}\omega}$—the turbulence scale. It can be expressed as:

$$l_{k\text{-}\omega} = k^{1/2}\beta_k \omega \tag{7}$$

where β_k—the model constant.

Based on RANS simulation, Herwig et al. [29] provided a way to quantitatively calculate the flow energy loss S_{pro} with 4 main sub-terms. $S_{\overline{pc}}$ and $S_{pc'}$ are the sub-term of entropy production caused by loss term, $S_{\overline{pd}}$ and $S_{pd'}$ are the sub-term of entropy production caused by dissipation term. They can be calculated by:

$$S_{\overline{pc}} = \frac{\lambda_t}{T^2}\left[\left(\frac{\partial \overline{T}}{\partial x}\right)^2 + \left(\frac{\partial \overline{T}}{\partial y}\right)^2 + \left(\frac{\partial \overline{T}}{\partial z}\right)^2\right] \tag{8}$$

$$S_{pc'} = \frac{\lambda_t}{T^2}\left[\overline{\left(\frac{\partial T'}{\partial x}\right)^2} + \overline{\left(\frac{\partial T'}{\partial y}\right)^2} + \overline{\left(\frac{\partial T'}{\partial z}\right)^2}\right] \tag{9}$$

$$S_{\overline{pd}} = \frac{\mu}{\overline{T}}\left[2\left(\frac{\partial \overline{u}}{\partial x}\right)^2 + 2\left(\frac{\partial \overline{v}}{\partial y}\right)^2 + 2\left(\frac{\partial \overline{w}}{\partial z}\right)^2 + \left(\frac{\partial \overline{u}}{\partial y} + \frac{\partial \overline{v}}{\partial x}\right)^2 + \left(\frac{\partial \overline{u}}{\partial z} + \frac{\partial \overline{w}}{\partial x}\right)^2 + \left(\frac{\partial \overline{v}}{\partial z} + \frac{\partial \overline{w}}{\partial y}\right)^2\right] \tag{10}$$

$$S_{pd} = \frac{\mu}{\overline{T}}\left[2\overline{\left(\frac{\partial u'}{\partial x}\right)^2} + 2\overline{\left(\frac{\partial v'}{\partial y}\right)^2} + 2\overline{\left(\frac{\partial w'}{\partial z}\right)^2} + \overline{\left(\frac{\partial u'}{\partial y} + \frac{\partial v'}{\partial x}\right)^2} + \overline{\left(\frac{\partial u'}{\partial z} + \frac{\partial w'}{\partial x}\right)^2} + \overline{\left(\frac{\partial v'}{\partial z} + \frac{\partial w'}{\partial y}\right)^2}\right] \tag{11}$$

where x, y, z—coordinate components. In the simulation of incompressible water flow in hydraulic turbomachineries, the sub-terms induced by velocity pulsation were dominant [30,31]. When using the SST k-ω model as turbulence model, the flow energy loss can be simplified to:

$$S_{pro} = \frac{\beta \rho \omega k}{T} \tag{12}$$

where β—the model constant of 0.09, k—the turbulent energy, ω—the turbulent eddy frequency, T—the temperature. With this method, the detail of flow energy loss inside pump turbine can be well visualized and analyzed based on CFD simulation results.

3.2. Computational Fluid Dynamics Setup

The numerical simulation was based on the commercial software ANSYS CFX. The flow passage of the pump turbine was from the draft tube inlet to the volute outlet as shown in Figure 2. The commercial software ICEMCFD was used to generate the grid. Tetrahedral and hexahedral grids were used in this study to ensure the computational quality and save computational cost. A grid independence check was conducted as shown in Figure 2a. The residual value of predicted head at the global best efficiency point under α = 14 degrees was chosen as the criterion and a residual less than 0.001 was accepted. The schematic diagram of the final grid is shown in Figure 2b. The grid number of each component is shown in Table 2. The total grid element number was 9,668,956.

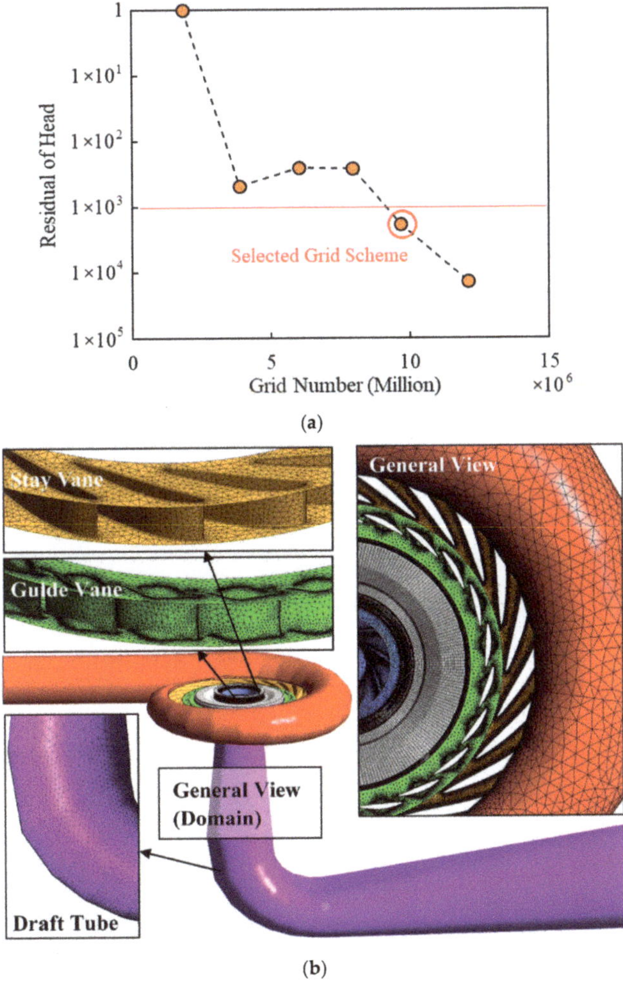

Figure 2. Fluid domain and grid. (**a**) Grid independence check; (**b**) the schematic map of the final grid.

Table 2. Grid number of each component.

Component	Grid Number
Draft tube	388,445
Runner (including clearance)	3,368,717
Guide vane	1,462,120
Stay vane	2,669,804
Volute	1,779,870
Total	9,668,956

A multi reference frame model was used in this study. The runner was set as the rotating domain, and the rest of the parts were stationary domains. The reference pressure was set to 1 Atm. The boundary conditions are given as follows. The inlet boundary was set at the inlet of draft tube with a given velocity. The outlet boundary was set at the volute outlet, and the given outlet static pressure was 0 Pa. The wall boundaries were no-slip wall type. The interfaces between rotor and stators were set as the stage type. The minimum number of iterations was set to 300, the maximum number of iterations was set to 1000 and the convergence criterion was RMS residual less than 1×10^{-5}.

3.3. Model Test

Model test is an important part of this study. As shown in Figure 3, the model scale pump turbine unit was installed in the test section. The upstream and downstream were storage tank and cavitation tank, respectively, which can play the role of water storage, rectification, joint regulation of cavitation with vacuum pump and so on. The supply pump was used to maintain the fluid circulation of the test system.

During the model test, the energy characteristics were measured at first. The parameters such as measured physical quantities and apparatus uncertainty are given in Table 3. Based on the constant speed test method, the working condition was changed by adjusting the valve and matching the guide vane opening angle, and the head H, flow rate Q, rotational speed n, shaft torque M and guide vane opening angle α were recorded. The values of power P and efficiency η can be calculated to determine the operating condition point of the unit. Among them, the head H was measured by the differential pressure sensor, and the static pressure p_s and the pressure difference p_{sd} between inlet and outlet were measured. The calculation formula is as follows:

$$H = \frac{p_{so} - p_{si}}{\rho g} = \frac{p_{sd}}{\rho g} \tag{13}$$

where ρ is density and g is the acceleration of gravity. Flow rate Q was measured by electro-magnetic flow meter. The measurement of power P needs to consider torque M and rotational speed n, where n was measured by rotating speed encoder and M was measured by torque meter. Power P can be calculated by:

$$P = \frac{2\pi n}{60} M \tag{14}$$

where the unit of n is [r/min]. The efficiency η can be calculated by:

$$\eta = \frac{\rho g Q H}{P} \tag{15}$$

The guide vane opening α was measured by using the angular displacement sensor. The above test parameters were transmitted to the data acquisition terminal for analog-to-digital conversion and post-processing.

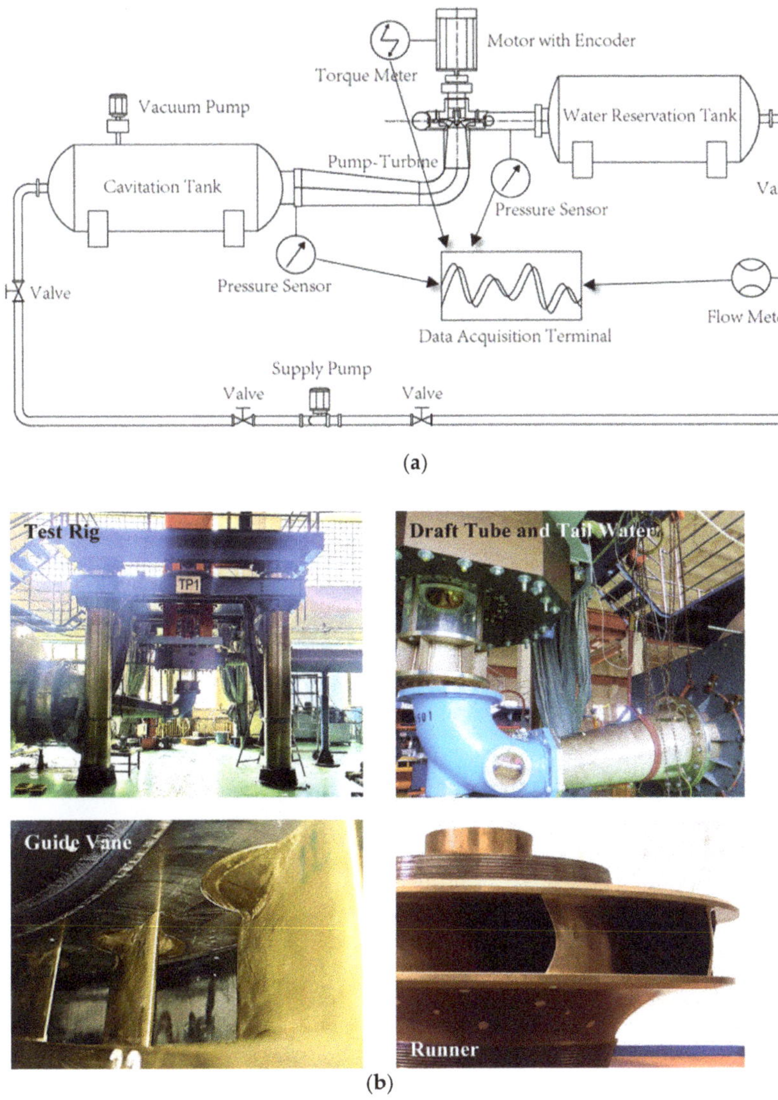

Figure 3. Test rig and apparatus. (**a**) The schematic map of the test rig and apparatus; (**b**) the on-site view of the test rig.

Table 3. Measured physical quantities and apparatus uncertainty.

Quantity	Apparatus	Type	Uncertainty
Flow rate	Electromagnetic flowmeter	Rosemount 8705TSE	±0.1%
Rotation speed	Rotary encoder	E6B2-CWZ1X	±0.02%
Head	Differential pressure sensor	SHAE 1151HP6E	±0.1%
Torque	Load sensor	GWT MP47/22C3	±0.015%
Tail-water pressure	Absolute pressure sensor	Rosemount 1151AP	±0.1%
Guide vane angle	Angular displacement sensor	BGJ 60	±0.10°

4. Comparison of Energy Performance

Figure 4 is the experimental–numerical comparison of the energy performance in pump mode. Five different guide vane opening angles were analyzed, including the head *H* and efficiency η. For the experimental data, the *Q*-*H* envelop curve and *Q*-η envelop curve are shown. The *Q*-η envelop curve is drawn based on the best efficiency point (Q_{BEP}) of each guide vane opening angle. The operation of the prototype pump turbine will follow this curve in order to have a higher efficiency. The head values *H* of these best efficiency points (Q_{BEP}) were recorded as the *Q*-*H* envelop curve. For each *Q*-*H* curve, its BEP point matches with the *Q*-η envelop curve and *Q*-*H* envelop curve. Generally, as shown by the *Q*-*H* envelop curve, *H* increases with a decrease in *Q*. The *Q*-η envelop curve has a peak (best efficiency point) as indicated in Figure 4. The global best efficiency point is under α = 14 degrees.

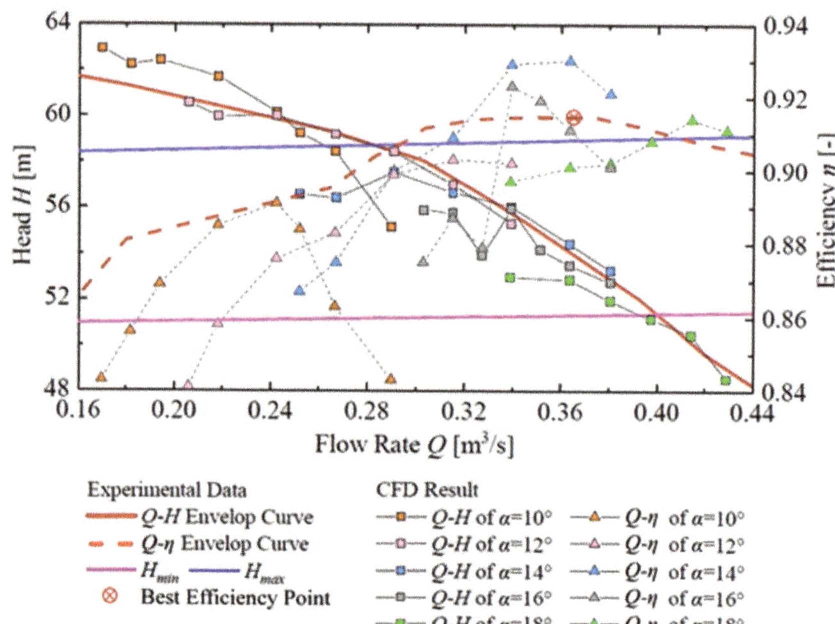

Figure 4. Experimental–numerical comparison of energy performance in pump mode.

In the analysis of CFD prediction results, the guide vane opening angle α from 18 degrees to 10 degrees was analyzed. With the decrease in the guide vane opening angle α, the flow rate becomes smaller and smaller. This shows the good flow-adjusting ability of the guide vane. The individual *Q*-*H* curves match the *Q*-*H* envelop curve well. Head *H* increases with the decrease in flow rate Q with local variations. For *Q*-η curves, each individual curve has a local peak. This means that the efficiency increases and decreases with the decrease in the flow rate. Each guide vane opening angle has a best operation range. The efficiencies of the CFD-predicted results are higher than the experimental data. This is reasonable because of the ignorance of the mechanical efficiency in the experiment. The individual *Q*-η curves have different laws because of the different guide vane openings. Detailed analyses are given in the following sections. CFD prediction will help the internal flow field analysis for the understanding of the relationship between the flow pattern and guide vane opening angle.

To have a better understanding of the drop in the head when the flow rate is relatively small, the head stagnation point and head valley point are defined for each α condition. For a *Q*-*H* curve, the slope is generally negative but one or multiple positive slope regions can be found. With a decrease in the flow rate, the point that H stops the increment is defined as the head stagnation point and the flow rate is denoted as Q_{STA}. If the flow rate continually decreases, H will rise again, the valley point is defined as the head valley point and the flow rate is denoted as Q_{VAL}. The values of Q_{STA} and Q_{VAL} are shown in Figure 5a,b. Both Q_{STA} and Q_{VAL} decrease with a decrease in α. This means that the unstable flow rate of the guide vane decreases synchronously with the decrease in the guide vane opening angle. A detailed analysis of the best efficiency η_{BEP} and flow rate of best efficiency points Q_{BEP} is shown in Figure 5c,d. The value of η_{BEP} increases within α = 10~14 degrees and decreases within α = 14~18 degrees. The value of Q_{BEP} generally increases from α = 10 degrees to α = 18 degrees, with an exception at α = 16 degrees. This means that the optimal matching opening angle of the guide vane increases synchronously with the increase in the flow rate. Therefore, the optimal matching and undesirable matching of the guide vane and impeller need further analyses.

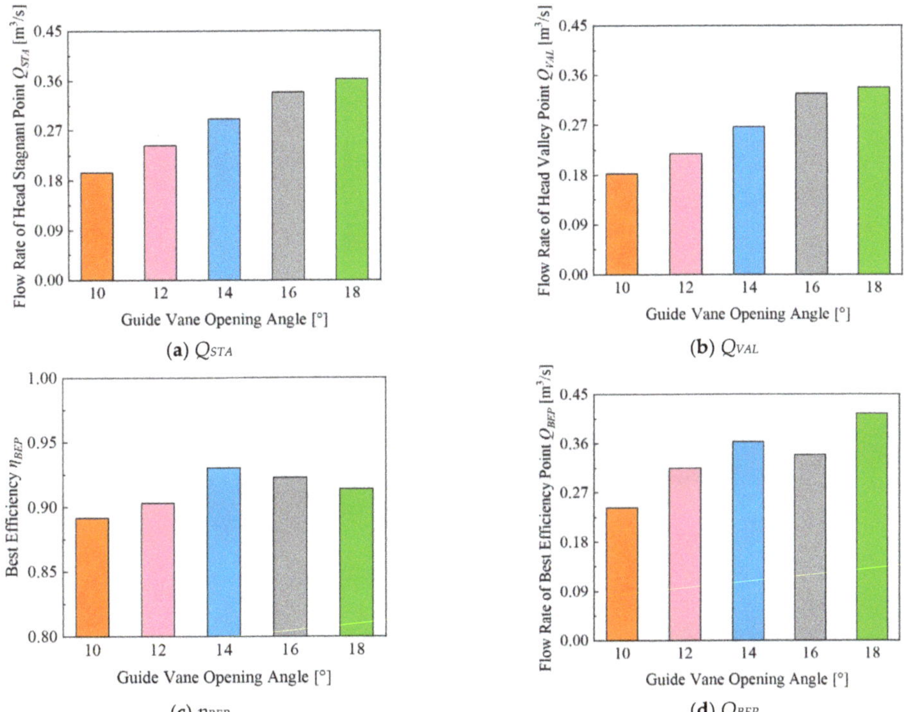

Figure 5. Variation in typical conditions of different attack angles α. (**a**) Q_{STA}. (**b**) Q_{VAL}. (**c**) η_{BEP}. (**d**) Q_{BEP}.

5. Analysis of Internal Flow

The analysis of the internal flow in the pump turbine includes five parts with five different guide vane opening angles. They are defined as the large guide vane opening angle (18 degrees), medium-large guide vane opening angle (16 degrees), medium guide vane opening angle (14 degrees), medium-small guide vane opening angle (12 degrees) and small guide vane opening angle (10 degrees), and are discussed in the following sections.

5.1. Large Guide Vane Opening Angle (18 Degrees)

Figure 6 shows the situation of the guide opening angle α of 18 degrees. Figure 6a, respectively, shows the flow–head *Q-H* and flow–efficiency *Q-η* Curve, where Q_{BEP} is approximately 0.415 m^3/s and Q_{VAL} is approximately 0.339 m^3/s. Through the analysis of the flow energy loss proportion inside the volute, stay vane, guide vane, runner and draft tube as shown in Figure 6b, it is found that, when the flow rate is at the Q_{BEP} point, the loss proportion of the guide vane is approximately 40.8%. The loss proportion inside the runner is close to that inside the guide vane. The loss proportion of the stay vane and volute is relatively small. The loss in the draft tube is less than 1%, which is very small. When the flow rate decreased to Q_{VAL}, the loss proportion of guide vane increased significantly to 53.9%, which was higher than the sum of other components. It can be seen from the S_{pro} contour and v vectors at the Q_{BEP} point that there is only a small amount of guide vane passages with an obvious loss increase, and that the intensity is not high. This is related to the local secondary flow structures. At the Q_{VAL} point, there are approximately seven channels of the guide vane with significant loss. The local intensity is very high, which is also related to the secondary flow in the guide vane channels, and it causes the flow blockage in the downstream components.

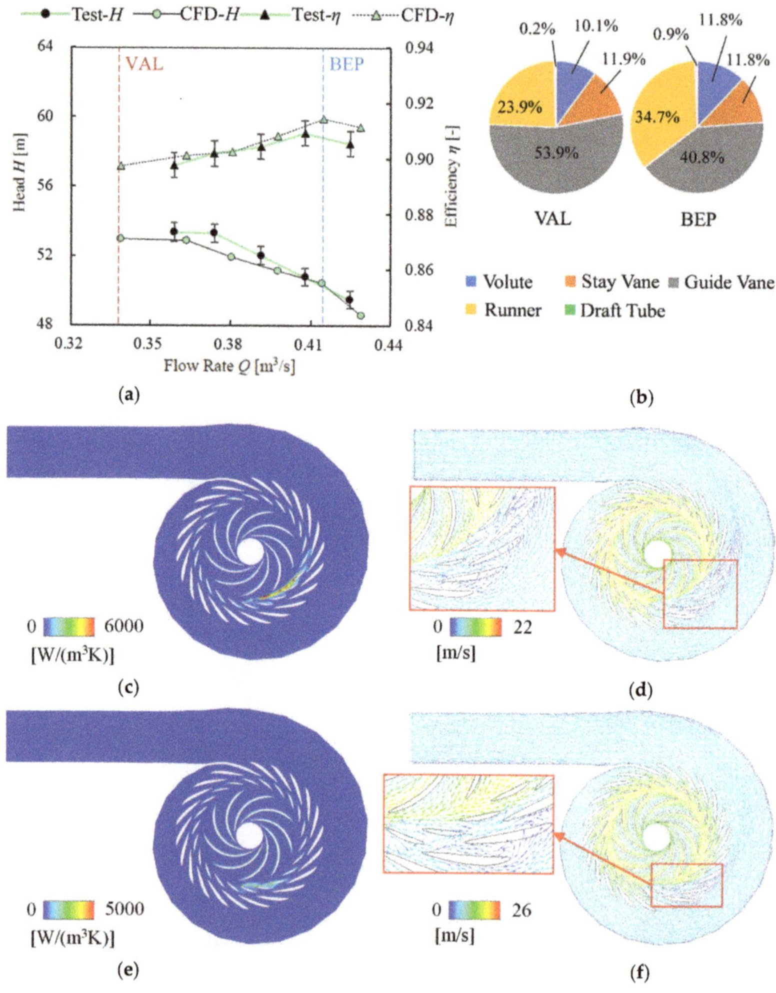

Figure 6. Performance curve and flow pattern at 18° attack angle. **(a)** *Q*-*H* and *Q*-η curve; **(b)** proportion of flow energy loss; **(c)** VAL S_{pro}; **(d)** VAL v; **(e)** BEP S_{pro}; **(f)** BEP v; VAL: local valley of *Q*-*H* curve; BEP: best efficiency point.

5.2. Medium-Large Guide Vane Opening Angle (16 Degrees)

Figure 7 shows the situation of the guide opening angle α of 16 degrees. Figure 7a, respectively, shows the flow–head Q-H and flow–efficiency Q-η curve, where Q_{BEP} is approximately 0.339 m^3/s and Q_{VAL} is approximately 0.327 m^3/s. As shown in Figure 7b, it is found that, when the flow rate is at the Q_{BEP} point, the loss proportion of the guide vane is approximately 51.8%, which is the largest. The loss proportion inside the runner is the second largest. The loss proportion of the stay vane and volute is smaller. The loss in the draft tube is also less than 1%, which is very small. When the flow rate decreased to Q_{VAL}, the loss proportion of the guide vane became 52.2%, which was almost unchanged. It can be seen from the S_{pro} contour and v vectors at the Q_{BEP} point and Q_{VAL} point that secondary flow structures are found in guide vane channels. Flow blockage in the downstream components can be also observed. Based on the comparison between the Q_{BEP} point and Q_{VAL} point, the intensity of the flow energy loss at the Q_{VAL} point is much stronger. This is why head H drops suddenly.

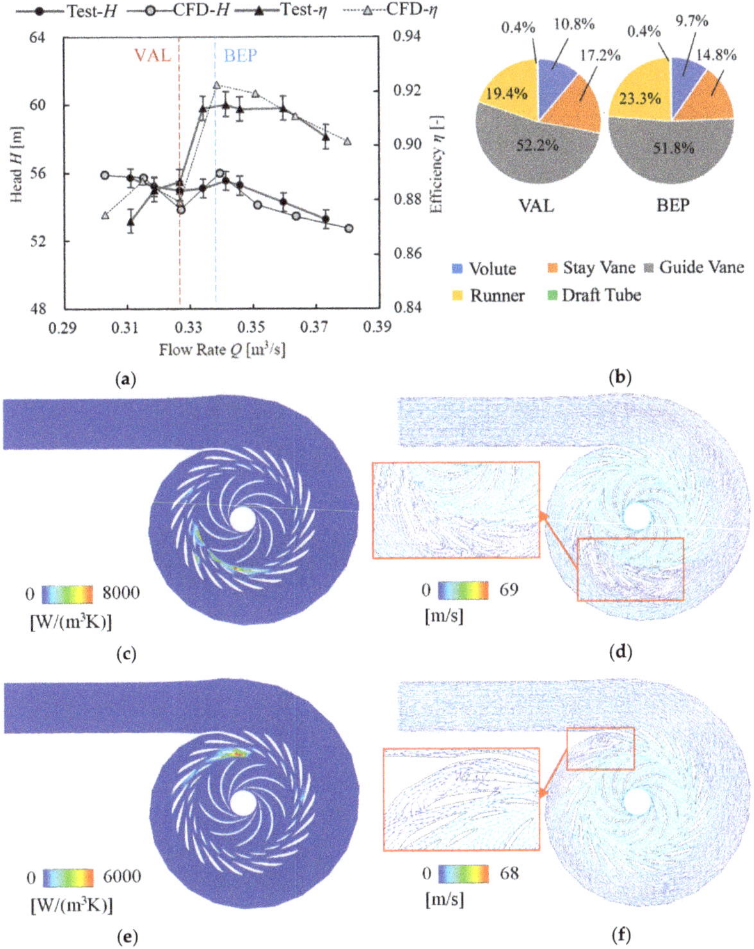

Figure 7. Performance curve and flow pattern at 16° attack angle. (**a**) Q-H and Q-η curve; (**b**) proportion of flow energy loss; (**c**) VAL S_{pro}; (**d**) VAL v; (**e**) BEP S_{pro}; (**f**) BEP v; VAL: local valley of Q-H curve; BEP: best efficiency point.

5.3. Medium Guide Vane Opening Angle (14 Degrees)

Figure 8 shows the situation of the guide opening angle α of 14 degrees. Figure 8a, respectively, shows the flow–head Q-H and flow–efficiency Q-η curve, where Q_{BEP} is approximately 0.363 m^3/s and Q_{VAL} is approximately 0.267 m^3/s. As shown in Figure 8b, it is found that when the flow rate is at the Q_{BEP} point, the loss proportion of the guide vane is approximately 36.9%. The loss proportion inside the runner is close to that inside the guide vane. The loss proportion of the stay vane and volute is smaller than that in the runner and guide vane. The loss in the draft tube is still much smaller. When the flow rate decreased to Q_{VAL}, the loss proportion of the guide vane increased significantly to 69.7%, which was higher than the sum of other components. The proportion of loss in the runner becomes lower. It can be seen from the S_{pro} contour and v vectors at the Q_{BEP} point that only few guide vane passages have a relatively strong energy loss and that the intensity is not that high. At the Q_{VAL} point, there are approximately eight channels of the guide vane with a significant flow energy loss with very high intensity. The sudden increase in S_{pro} also indicates the mismatching of the guide vane and runner incoming flow. The bad flow regime will cause flow blockage in the stay vane and volute.

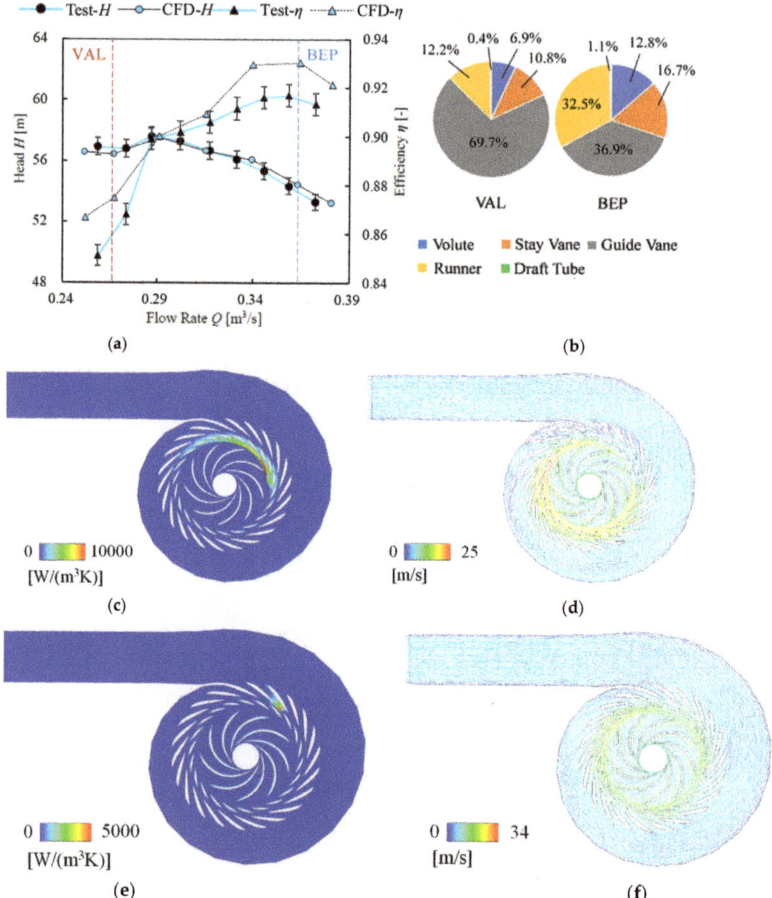

Figure 8. Performance curve and flow pattern at 14° attack angle. (**a**) Q-H and Q-η curve; (**b**) proportion of flow energy loss; (**c**) VAL S_{pro}; (**d**) VAL v; (**e**) BEP S_{pro}; (**f**) BEP v; VAL: local valley of Q-H curve; BEP: best efficiency point.

5.4. Medium-Small Guide Vane Opening Angle (12 Degrees)

Figure 9 shows the situation of the guide opening angle α of 12 degrees. Figure 9a, respectively, shows the flow–head Q-H and flow–efficiency Q-η curve, where Q_{BEP} is approximately 0.315 m^3/s and Q_{VAL} is approximately 0.219 m^3/s. As shown in Figure 9b, it is found that, when the flow rate is at the Q_{BEP} point, the loss proportion of the guide vane is approximately 47.5%, which is the largest. The loss proportion inside the runner, stay vane and volute is similar. The loss in the draft tube is still less than 1%. When the flow rate decreased to Q_{VAL}, the loss proportion of the guide vane increased significantly to 65.6%, which was higher than the sum of other components. The proportion of loss in the volute becomes obviously lower. It can be seen from the S_{pro} contour and v vectors at the Q_{BEP} point that the loss is relatively low and the flow regime is smooth in all of the guide vane channels. At the Q_{VAL} point, some guide vane channels have a high flow energy loss with a high intensity. This is related to the secondary flow in the guide vane, and the downstream components are influenced.

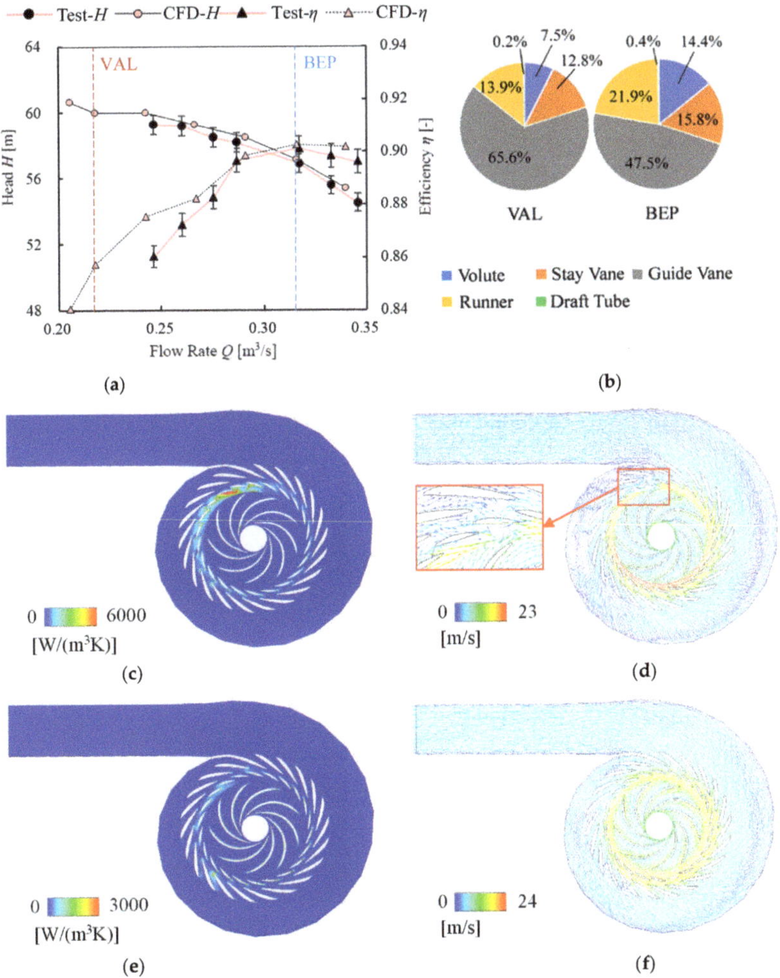

Figure 9. Performance curve and flow pattern at 12° attack angle. (**a**) Q-H and Q-η curve; (**b**) proportion of flow energy loss; (**c**) VAL S_{pro}; (**d**) VAL v; (**e**) BEP S_{pro}; (**f**) BEP v; VAL: local valley of Q-H curve; BEP: best efficiency point.

5.5. Small Guide Vane Opening Angle (10 Degrees)

Figure 10 shows the situation of the guide opening angle α of 10 degrees. Figure 10a, respectively, shows the flow–head Q-H and flow–efficiency Q-η curve, where Q_{BEP} is approximately 0.243 m^3/s and Q_{VAL} is approximately 0.183 m^3/s. As shown in Figure 10b, it is found that, when the flow rate is at the Q_{BEP} point, the loss proportion of the guide vane is approximately 54.8%, which is the largest. The loss proportion inside the runner and stay vane is similar and lower than that in the guide vane. The loss in the volute is much lower. The loss in the draft tube is less than 1%, which is very low. When the flow rate decreased to Q_{VAL}, the loss proportion of the guide vane increased significantly to 70.5%, which was higher than the sum of other components and completely dominant. The proportion of loss in the stay vane and volute becomes obviously lower. It can be seen from the S_{pro} contour and v vectors at Q_{BEP} point that the loss is relatively low and the flow regime is smooth in all of the guide vane channels. At the Q_{VAL} point, the flow regime in the guide vane becomes bad and causes an increase in the flow energy loss.

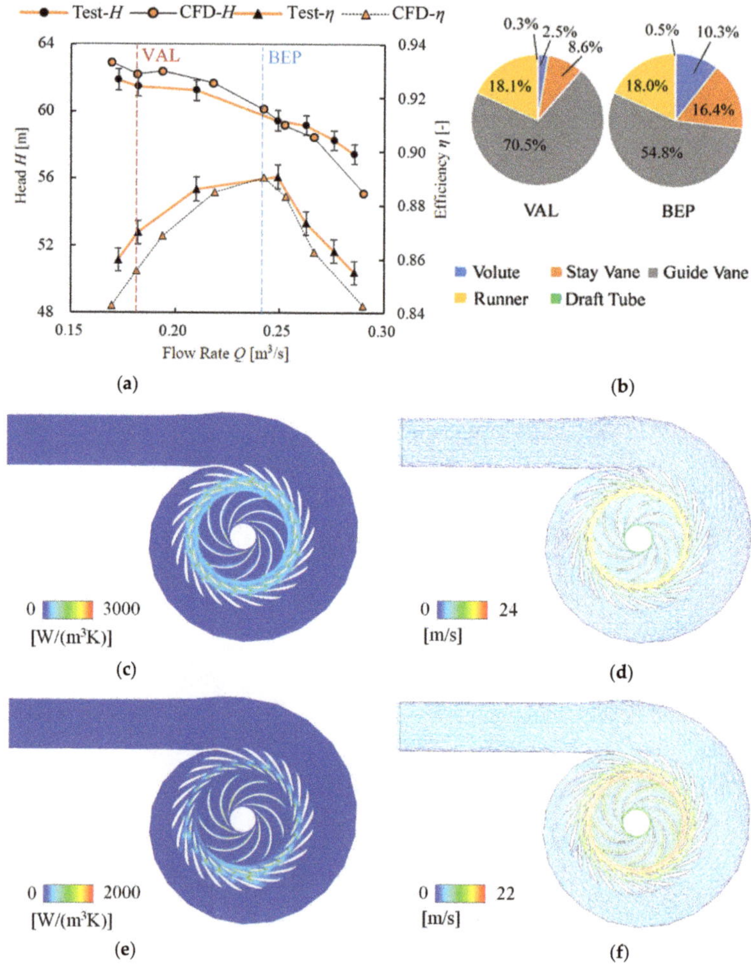

Figure 10. Performance curve and flow pattern at 10° attack angle. **(a)** Q-H and Q-η curve; **(b)** proportion of flow energy loss; **(c)** VAL S_{pro}; **(d)** VAL v; **(e)** BEP S_{pro}; **(f)** BEP v; VAL: local valley of Q-H curve; BEP: best efficiency point.

6. Discussion

The stagnation and drop in head H with the decrease in flow rate Q is related to the mismatching of the guide vane flow angle. In order to have a better understanding of this mis-matching, Figure 11 is schematically drawn with two conditions of Q_{BEP} and Q_{VAL}. The most important change from the Q_{BEP} point to Q_{VAL} point is the decrease in the flow rate.

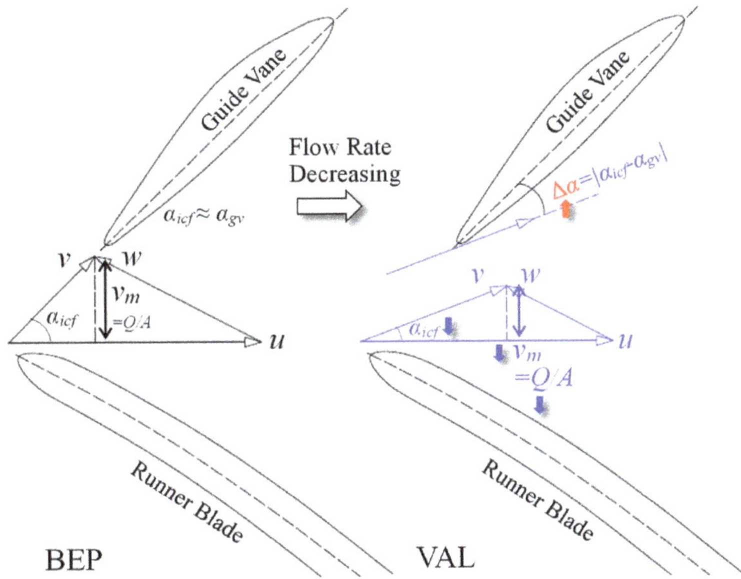

Figure 11. Schematic map of the mismatching of guide vane flow angle.

The most important analysis method of turbomachinery is the velocity triangle analysis. If we draw the velocity triangle at the runner trailing-edge in pump mode, the rotational linear velocity u can be determined at a specific rotational speed. The direction of relative velocity w can be determined because the runner blade is fixed. The meridional component of absolute velocity v_m can be calculated by [32]:

$$v_m = \frac{Q}{A} \tag{16}$$

where A is the flow passing area where v_m is analyzed.

When the flow rate is Q_{BEP}, which represents a good flow condition, the absolute flow angle between the runner and guide vane, defined as α_{icf}, will be almost equal to the guide vane installation angle α_{gv}. When the flow rate Q decreases, v_m will become smaller. Because the direction of w is almost unchanged, α_{icf} will be smaller, as indicated. The absolute difference $\Delta\alpha$ between α_{gv} and α_{icf} can be defined as:

$$\Delta\alpha = \left|\alpha_{icf} - \alpha_{gv}\right| \tag{17}$$

Therefore, $\Delta\alpha$ will be larger. This means that the good matching between the flow angle and guide vane installation angle will be broken when the flow rate decreases. If this mismatching becomes strong, the flow regime in guide vane channels will be very bad, leading to a 'stall' situation. This is why the stagnation of and drop in head H occur. This mis-matching will also occur when the flow rate increases. This is why the efficiency curve has an optimal region. For the large flow rate conditions, there is no positive Q-H slope problem.

In addition, the slip phenomenon of the blade channel caused by the blade number limitation will also make the originally estimated incoming flow direction upstream to the guide vane deflect to some extent. This phenomenon is not the same in different turbomachinery [33]. It will be strong in a pump turbine and will cause a change in velocity triangle components. As for the effect of the slip phenomenon on the *Q-H* instability of the pump turbine, we will pay special attention to it in future research. In general, the mismatching of the guide vane flow angle is especially important in small flow rate conditions, which is extremely off-design.

7. Conclusions

In this study, the flow energy loss and *Q-H* stability in a reversible pump turbine in pump mode was discussed for five different guide vane opening angles. These opening angle conditions cover the main conditions during daily operation in pump mode. The mismatching of the guide vane and runner incoming flow will induce an increase in flow energy loss in the guide vane and downstream components. It will cause a drop in head *H* and instability in the *Q-H* relationship. This situation occurs at both the large guide vane opening and small guide vane opening. With a decrease in the flow rate, the angle difference between the runner outlet flow and guide vane becomes bigger and bigger in order to generate the positive slope *Q-H* region. If this positive slope region overlaps a specific head, the operation stability will be influenced.

In the current pump turbine operation, the selection of the on-cam operation curve depends on the judgment of the high-efficiency envelop. However, the *Q-H* unstable region of the pump turbine in pump mode is also very important. As discussed in this study, if only considering high efficiency, the unit may face the risk of an unstable operation. Based on the research results of this paper, we can determine a scheme of selecting the on-cam envelop curve that takes into account both efficiency and stability without changing the hydraulic design. Our future research will compare and discuss the advantages and disadvantages of the selection of the on-cam operation scheme.

Generally, this study provides a reference for the solution of engineering problems, mainly serving the stable operation of the unit. This study is of great significance for the technical improvement in large-capacity medium and high head pumped storage power stations.

Author Contributions: Methodology, Z.W.; software, W.Y.; validation, R.T.; formal analysis, W.Y.; investigation, D.Z.; resources, W.Y.; writing—original draft preparation, W.Y.; writing—review and editing, R.T.; visualization, D.Z.; supervision, Z.W.; project administration, W.Y.; funding acquisition, R.T. All authors have read and agreed to the published version of the manuscript.

Funding: This research was funded by the Open Research Fund Program of State Key Laboratory of Hydroscience and Engineering (No. sklhse-2022-E-01).

Institutional Review Board Statement: Not applicable.

Informed Consent Statement: Not applicable.

Data Availability Statement: Not applicable.

Conflicts of Interest: The authors declare no conflict of interest.

References

1. Jin, B. Research on performance evaluation of green supply chain of automobile enterprises under the background of carbon peak and carbon neutralization. *Energy Rep.* **2021**, *7*, 594–604. [CrossRef]
2. Tan, X.; Liu, J.; Xu, Z.; Yao, L.; Ji, G.; Shan, B. Power supply and demand balance during the 14th five-year plan period under the goal of carbon emission peak and carbon neutrality. *Electr. Power* **2021**, *54*, 1–6.
3. Kong, Y.; Kong, Z.; Liu, Z.; Wei, C.; Zhang, J.; An, G. Pumped storage power stations in China: The past, the present, and the future. *Renew. Sustain. Energy Rev.* **2017**, *71*, 720–731. [CrossRef]
4. He, Y.; Guan, L.; Cai, Q.; Liu, X.; Li, C. Analysis of securing function and economic benefit of pumped storage station in power grid. *Power Syst. Technol.* **2004**, *28*, 54–57+67.

5. Xu, F.; Chen, L.; Jin, H.; Liu, Z. Modeling and application analysis of optimal joint operation of pumped storage power station and wind power. *Autom. Electr. Power Syst.* **2013**, *37*, 149–154.
6. Anagnostopoulos, J.S.; Papantonis, D.E. Pumping station design for a pumped-storage wind-hydro power plant. *Energy Convers. Manag.* **2007**, *48*, 3009–3017. [CrossRef]
7. Yan, J.; Koutnik, J.; Seidel, U.; Hubner, B. Compressible simulation of rotor-stator interaction in pump-turbines. *IOP Conf. Ser.-Earth Environ. Sci.* **2010**, *12*, 012008. [CrossRef]
8. Li, D.; Gong, R.; Wang, H.; Wei, X.; Liu, Z.; Qin, D. Unstable head-flow characteristics of pump-turbine under different guide vane openings in pump mode. *J. Drain. Irrig. Mach. Eng.* **2016**, *34*, 1–8.
9. Zhu, D.; Xiao, R.; Tao, R.; Liu, W. Impact of guide vane opening angle on the flow stability in a pump-turbine in pump mode. *Proc. Inst. Mech. Eng. Part C-J. Mech. Eng. Sci.* **2017**, *231*, 2484–2492. [CrossRef]
10. Yao, Y.; Xiao, Y.; Zhu, W.; Zhai, L.; An, S.; Wang, Z. Numerical analysis of a model pump-turbine internal flow behavior in pump hump district. *IOP Conf. Ser.-Earth Environ. Sci.* **2014**, *22*, 032040.
11. Li, D.; Wang, H.; Xiang, G.; Gong, R.; Wei, X.; Liu, Z. Unsteady simulation and analysis for hump characteristics of a pump turbine model. *Renew. Energy* **2015**, *77*, 32–42.
12. Li, D.; Wang, H.; Qin, Y.; Wei, X.; Qin, D. Numerical simulation of hysteresis characteristic in the hump region of a pump-turbine model. *Renew. Energy* **2018**, *115*, 433–447. [CrossRef]
13. Xiao, Y.; Yao, Y.; Wang, Z.; Zhang, J.; Luo, Y.; Zeng, C.; Zhu, W. Hydrodynamic mechanism analysis of the pump hump district for a pump-turbine. *Eng. Comput.* **2016**, *33*, 957–976. [CrossRef]
14. Lu, Z.; Xiao, R.; Tao, R.; Li, P.; Liu, W. Influence of guide vane profile on the flow energy dissipation in a reversible pump-turbine at pump mode. *J. Energy Storage* **2022**, *49*, 104161. [CrossRef]
15. Song, H.; Zhang, J.; Huang, P.; Cai, H.; Cao, P.; Hu, B. Analysis of rotor-stator interaction of a pump-turbine with splitter blades in a pump mode. *Mathematics* **2020**, *8*, 1465. [CrossRef]
16. Zhang, Y.; Chen, T.; Li, J.; Yu, J. Experimental study of load variations on pressure fluctuations in a prototype reversible pump turbine in generating mode. *J. Fluids Eng.* **2017**, *139*, 074501. [CrossRef]
17. Zhang, C.; Xia, L.; Diao, W. Influence of flow structures evolution on hump characteristics of a model pump-turbine in pump mode. *J. Zhejiang Univ. Eng. Sci.* **2017**, *51*, 2249–2258.
18. Yang, J.; Pavesi, G.; Liu, X.; Xie, T.; Liu, J. Unsteady flow characteristics regarding hump instability in the first stage of a multistage pump-turbine in pump mode. *Renew. Energy* **2018**, *127*, 377–385. [CrossRef]
19. Wang, H.; Wu, G.; Wu, W.; Wei, X.; Chen, Y.; Li, H. Numerical simulation and analysis of the hump district of Francis pump-turbine. *J. Hydroelectr. Eng.* **2012**, *31*, 253–258.
20. Yang, J.; Yuan, S.; Pavesi, G.; Li, C.; Ye, Z. Study of hump instability phenomena in pump turbine at large partial flow conditions on pump mode. *J. Mech. Eng.* **2016**, *52*, 170–178. [CrossRef]
21. Zhao, W.; Wu, B.; Xu, J. Aerodynamic design and analysis of a multistage vaneless counter-rotating turbine. *J. Turbomach.* **2014**, *137*, 061008. [CrossRef]
22. Vanzante, D.E.; To, W.M.; Chen, J.P. Blade row interaction effects on the performance of a moderately loaded NASA transonic compressor stage. *Am. Soc. Mech. Eng.* **2003**, *1*, 969–980.
23. Sun, Y.; Ren, Y. Unsteady loss analyses of the flow in single-stage transonic compressor. *J. Tsinghua Univ. Sci. Technol.* **2009**, *49*, 759–762.
24. Esfahani, J.A.; Modirkhazeni, M.S. Accuracy analysis of predicted velocity profiles of laminar duct flow with entropy generation method. *Appl. Math. Mech.* **2013**, *34*, 971–984. [CrossRef]
25. Kluxen, R.; Behre, S.; Jeschke, P.; Guendogdu, Y. Loss mechanisms of interplatform steps in a 1.5-stage axial flow turbine. *J. Turbomach.* **2017**, *139*, 031007. [CrossRef]
26. Soltanmohamadi, R.; Lakzian, E. Improved design of Wells turbine for wave energy conversion using entropy generation. *Meccanica* **2016**, *51*, 1713–1722. [CrossRef]
27. Zeinalpour, M.; Mazaheri, K. Entropy minimization in turbine cascade using continuous adjoint formulation. *Eng. Optim.* **2015**, *48*, 213–230. [CrossRef]
28. Menter, F.R.; Kuntz, M.; Langtry, R. Ten years of industrial experience with the SST turbulence model. *Turbul. Heat Mass Transf.* **2003**, *4*, 625–632.
29. Herwig, H.; Kock, F. Direct and indirect methods of calculating entropy generation rates in turbulent convective heat transfer problems. *Heat Mass Transf.* **2007**, *43*, 207–215. [CrossRef]
30. Tao, R.; Zhao, X.; Wang, Z. Evaluating the transient energy dissipation in a centrifugal impeller under rotor-stator interaction. *Entropy* **2019**, *21*, 271. [CrossRef]
31. Zhou, Q.; Xia, L.; Zhang, C. Internal mechanism and improvement criteria for the runaway oscillation stability of a pump-turbine. *Appl. Sci.* **2018**, *8*, 2193. [CrossRef]
32. Guan, X. *Modern Pumps Theory and Design*; China Astronautic Publishing House: Beijing, China, 2011; pp. 32–37.
33. Capurso, T.; Stefanizzi, M.; Pascazio, G.; Camporeale, S.M.; Torresi, M. Dependency of the slip phenomenon on the inertial forces inside radial runners. *AIP Conf. Proc.* **2019**, *2191*, 020034.

Article

Improvement of the Flow Pattern of a Forebay with a Side-Intake Pumping Station by Diversion Piers Based on Orthogonal Test Method

Chen Zhang [1], Haodi Yan [2], Muhammad Tahir Jamil [2,*] and Yonghai Yu [2]

[1] College of Mechanics and Materials, Hohai University, Nanjing 210098, China
[2] College of Agricultural Science and Engineering, Hohai University, Nanjing 210098, China
* Correspondence: tahir30@yahoo.com

Abstract: The flow analysis of the forebay of a lateral intake pumping station with asymmetrical operating pumps was carried out with a realizable k-ε turbulent model and SIMPLEC (Semi Implicit Method for Pressure Linked Equations Consistent) algorithm. The Pressure Inlet boundary condition was adopted and the pressure between the top surface and the bottom surface was linear with the height of the inlet section. The Mass Flow Outlet boundary condition was also adopted to ensure the accuracy and precision of the CFD (Computational Fluid Dynamics) simulation. The diversion pier was selected as the optimization strategy based on the flow parameters. The layout of the diversion piers was designed with four parameters which are the relative length, relative height, width, and straight-line distance of the piers' tail. Each parameter had three values. Based on the orthogonal test, nine groups of the numerical simulation on different layouts of diversion piers were analyzed with the uniformity of axial flow velocity and weighted average angle of the flow velocity of the inlet cross-section of each pump, reducing the number of tests from 64 (4^3) groups to 9 groups, improving work efficiency. The results show that the diversion piers had a significant adjustment of uniformity of axial flow velocity and weighted average angle of flow velocity. After optimization of the forebay, the uniformity of axial flow velocity of intake of No.1 pump was 80.26% and the weighted average angle of flow velocity was 77.68°. The above values of the No.2 pump were 98.74% and 87.84°, respectively. The values of the No.4 pump were 93.41% and 77.28°. The results of numerical simulation, which was carried out to estimate the rectification effect under the operation combination of the No.1, No.3, and No.4 pumps, showed that the uniformity and the angle of the No.1 pump were 92.65% and 72.66°, respectively, the uniformity and the angle of No.3 pump were 94.54% and 85.14°, and the uniformity and the angle of the No.4 pump were 75.81% and 78.21°. This research proves that the orthogonal test method, in a reasonable and convenient way, can be applied in hydraulic optimization for a lateral intake pumping station.

Keywords: flow pattern; orthogonal test method; lateral intake; CFD numerical simulation; diversion pier

Citation: Zhang, C.; Yan, H.; Jamil, M.T.; Yu, Y. Improvement of the Flow Pattern of a Forebay with a Side-Intake Pumping Station by Diversion Piers Based on Orthogonal Test Method. *Water* **2022**, *14*, 2663. https://doi.org/10.3390/w14172663

Academic Editor: Anargiros I. Delis

Received: 30 July 2022
Accepted: 26 August 2022
Published: 28 August 2022

Publisher's Note: MDPI stays neutral with regard to jurisdictional claims in published maps and institutional affiliations.

Copyright: © 2022 by the authors. Licensee MDPI, Basel, Switzerland. This article is an open access article distributed under the terms and conditions of the Creative Commons Attribution (CC BY) license (https://creativecommons.org/licenses/by/4.0/).

1. Introduction

In the forebay of a lateral intake pumping station, vortices, spiral flows, and large-scale reversed flows occurred, directly affecting the pump's inlet conditions, and resulting in a decline in the pump's energy performance and steam erosion performance. These flow patterns also cause vibration in the unit and may even jeopardize its safety. Therefore, rectification measures must be developed for the lateral intake pumping stations [1–3]. In order to improve flow performance, some researchers have studied three-dimensional flows in pump or pumping station intakes [4–8]. Several studies have been conducted on the flow and the alteration of the flow pattern in the forebay of the pumping stations. Kadam P. et al. [9] studied the flow field of the inlet building of a pumping station using physical model tests along with numerical simulation techniques. They also evaluated the flow characteristics of the forebay and the inlet pipe and discovered that the poor intake flow pattern was caused by the high diffusion angle of the forebay and small inundation depth. To improve the flow in the forebay of the pumping station, Liu, C. et al. [10,11] used division piers that study and validate conclusions numerically and experimentally. Feng X. [12] used the rectifier sill as a rectifying measure for the forward inflow forebay of the pumping station. Li J. et al. [13] simulated flow patterns in the forebay and suction sump of the Tianshan pumping station through a finite element analysis approach. Xu, B. et al. [14] studied the influence of the length of the diversion piers on the flow pattern in an asymmetric combined sluice-pump station project based on the CFD numerical simulation. Finally, Xia, C. et al. [15] added the inverted T-shaped diversion piers in the forebay to significantly improve the uniformity of the internal flow velocity distribution.

Luo, C. et al. [16] added the sill to rectify the flow pattern in the forebay and analyzed the general rules of the position and height of the sill. Yu, Y. et al. [17] studied the influence of the flow deflector on the flow pattern in the forebay of the pumping station and obtained the reasonable layout parameters of the flow deflector based on the physical model test and numerical simulation. According to a study by Luo, C. et al. [18], opening the diversion pier can lower the lateral and axial flow velocity as well as the oblique flow area of the surface layer of the flow surface in front of the pier head, enhancing navigational safety. The orthogonal test method is a scientific approach to testing a design based on the analysis of multi-component and multi-level test situations. It significantly reduces the number of tests while allowing for the investigation of the influence of each factor level on evaluation indices through the creation of orthogonal test tables and ranges. The influence of each factor on the head of the pump was investigated by Wang, W. et al. [19] using an orthogonal test design with four factors, three levels, and numerical simulation calculation. J. Zhou [20] obtained the influence of each factor on the uniformity of axial flow velocity and the weighted average angle of velocity by the length, width, radian, and relative height of the orthogonal test design of five factors and four levels. The study by Xu, B. et al. [21] showed that the orthogonal test method has a significant effect on the parameter design for the rectification measures of pumping stations.

The orthogonal test method can optimize the hydraulic design of the diversion piers. Numerous studies combining physical model testing and numerical simulation have been carried out for the rectification measures for the forebay of pumping stations. These studies show that the results of the CFD numerical simulation are essentially similar to those of the physical model testing and that the only approach to enhance the flow pattern of the forebay of pumping stations is through the use of the CFD numerical simulation method [22–26]. As a result, in this paper, the CFD numerical simulation method will be used to analyze the flow pattern of the forebay of a specific pumping station and to improve the design parameters of the diversion pier. The analysis method and procedure are shown in Figure 1.

The novelty of this study lies in the object of flow analysis, which includes not only the forebay with side-intake of the pumping station but also asymmetrical operating pumps and nine groups of numerical simulation on the various layouts of diversion piers investigated using the orthogonal test with uniform axial flow velocity and weighted average angle of the flow velocity at the inlet cross-section of each pump. This study is divided into five sections: study area and method, typical sections in calculation domain and evaluation indexes of flow pattern in the forebay, design of diversion piers based on the orthogonal test, CFD analysis of Pumps 1, 3, and 4, and conclusion. This study provides valuable resources for related topics of interest in pumping station engineering.

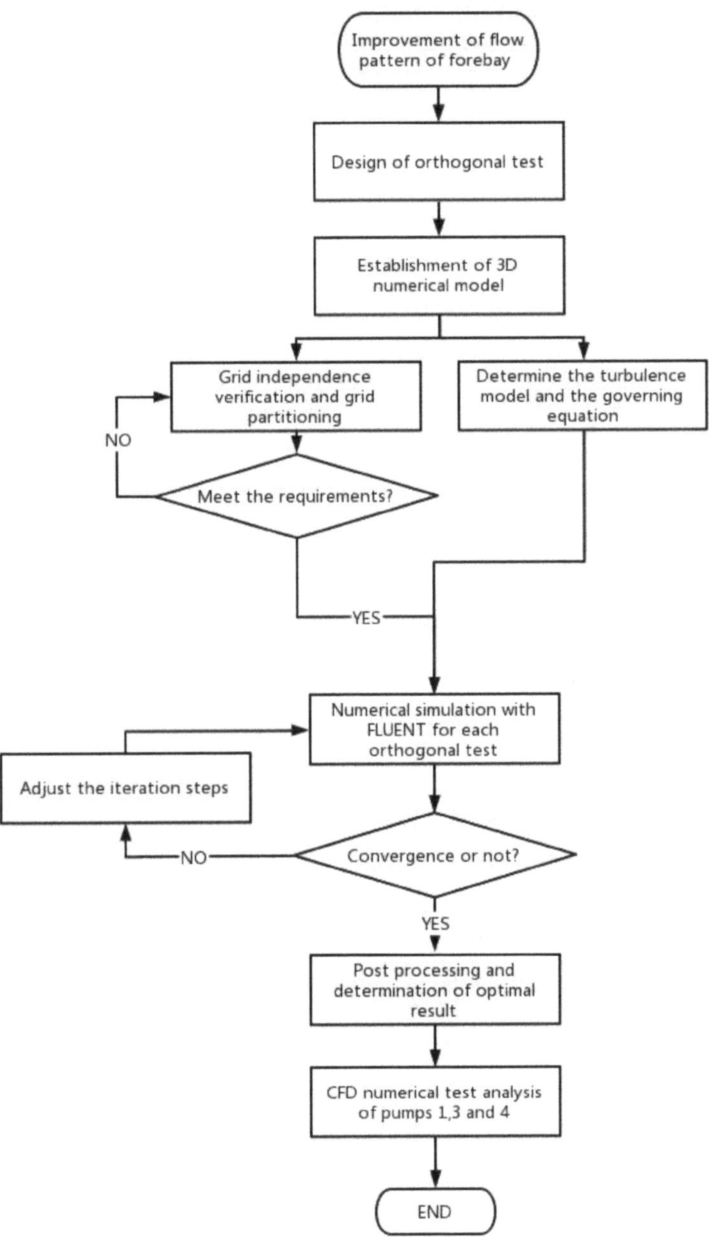

Figure 1. Analysis flowchart.

2. Study Area and Method

2.1. Study Area

There are four pumps in the certain lateral inlet pumping station in Shandong province, China, among which, the rated flow of the No.1 or No.4 pump is 2.6 m^3/s and the rated flow of the No.2 or No.3 pump is 5.4 m^3/s. The inlet cross-section of every pump is 3.2 m high and 4.85 m wide. The water depth of the forebay before the slope is 4 m, the water depth of the forebay behind the slope is 5.4 m, and the bottom slope has a gradient of 1:8, as shown in Figure 2.

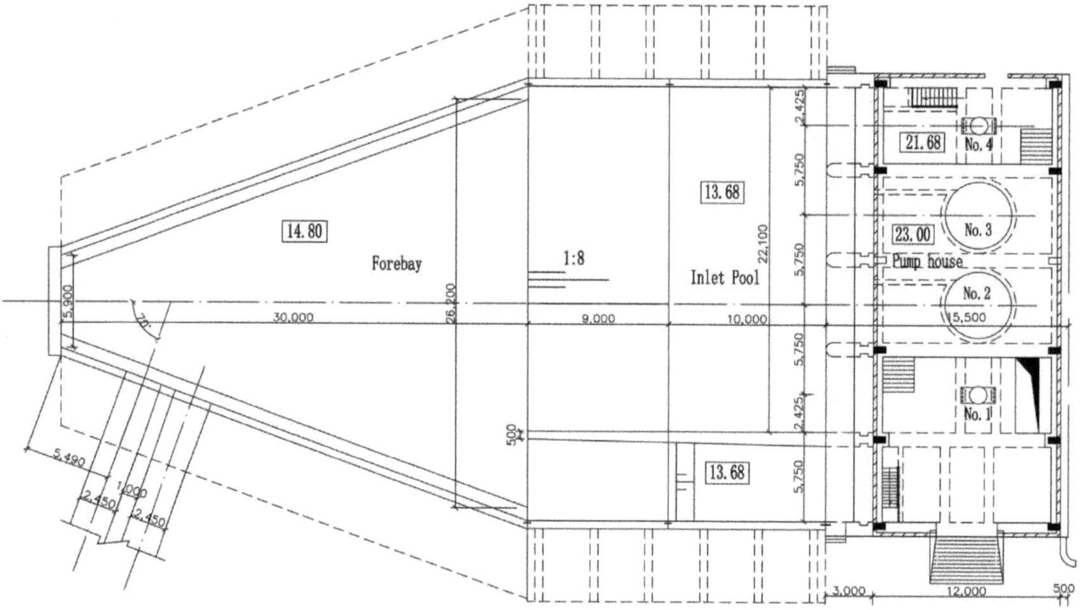

Figure 2. The plane layout of the certain lateral intake pumping station (EL: m, OTH: mm:). (Scale 1:200).

Two main pump operation schemes are the combination of No.1 pump, No.2 pump, and No.4 pump and the combination of No.1 pump, No.3 pump, and No.4 pump. The rectification measures with diversion piers of the forebay under the combination of pumps 1, 2, and 4 will be mainly studied. The effectiveness of the rectification measures of pumps 1, 3, and 4 will be tested.

2.2. Study Method

FLUENT is used to numerically simulate the flow pattern of the forebay. The flow governing equations are the continuity equation and the Navier-Stokes equation. LES (Large eddy simulation) and RANS (Reynolds averaged Navier-Stokes) are two simulation methods of turbulence. LES can capture large-scale effects and coherent structures in unsteady and nonequilibrium processes and RANS will be incapable of action, but the computation load will be bigger, and consumption time will be longer for LES. The simulation accuracy by RANS can meet the requirement. The realizable k-ε turbulence model for RANS was suitable for the simulation of the flow in the forebay with side-intake of the pumping station [3]. The Reynolds number in the forebay is 2.86×10^5 greater than 3×10^4, which shows that the turbulence is fully developed. The SIMPLEC algorithm is adopted [27].

2.2.1. Boundary Conditions

The side inlet cross-section is used as the inlet of the calculation domain, and the pressure inlet boundary condition was adopted. The pressure on the top of the inlet section is 8 kPa. The pressure at the bottom of the inlet section is 40 kPa. The pressure between the top surface and the bottom surface is linear with the height of the inlet section. For the free surface of the forebay, the shear stress generated by the air on the water surface and heat exchange can be ignored, and the treatment of the symmetrical boundary with the surface of the forebay was adopted according to the rigid-lid assumption. The outlet of the discharge pipe of every pump was used as the outlet boundary. The Mass Flow Outlet boundary condition was then adopted. The mass flow of the No.1 pump was 2600 kg/s, the mass flow of the No.2 pump was 5400 kg/s, and the mass flow of the No.4 pump was 2600 kg/s. All the solid parts were set to the wall according to the wall function method.

2.2.2. Grid Independence Verification

The computing domain is meshed by Fluent Meshing, including a tetrahedral grid in complex parts and hexahedral mesh for non-complex parts. In order to improve the calculation speed and accuracy of the computational domain, grid independence verification should be done. The number of mesh is selected from 180,000, 320,000, 570,000, 800,000, 1.41 million, 3.33 million, and 4.93 million. Head loss, h_f, was used as an evaluation indicator for grid independence verification.

$$h_f = \frac{(P_{in} - P_{out})}{\rho g} \quad (1)$$

where P_{in} is the total pressure of the inlet; P_{out} is the total pressure of the outlet; ρ is the density of water; and g is the gravity acceleration.

According to Figure 3, after the number of mesh exceeds 800,000, the head loss of the calculation domain changes little and its head loss does not exceed 2%. Therefore, the number of mesh was chosen as 800,000.

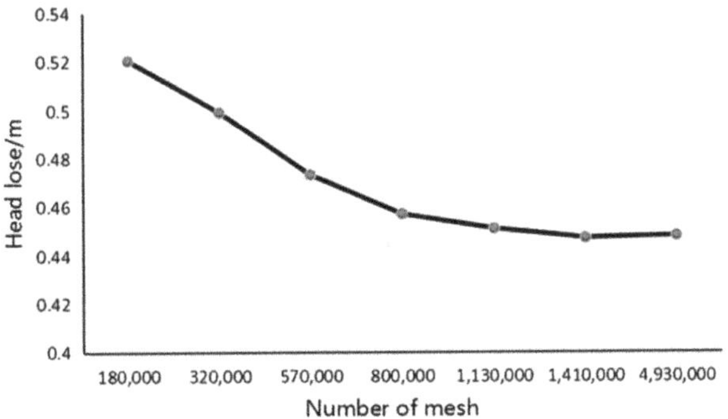

Figure 3. Head loss at different numbers of mesh.

3. Typical Sections in Calculation Domain and Evaluation Indexes of Flow Pattern in Forebay

As shown in Figure 4, three horizontal sections are taken in the upper, middle, and lower layers of the calculation domain as typical sections, which are called horizontal section I (Z = 1.4 m), horizontal section II (Z = 0.1 m), and horizontal section III (Z = −1 m). The elevation of the bottom of the forebay, as indicated in Figure 1, is Z = 0.

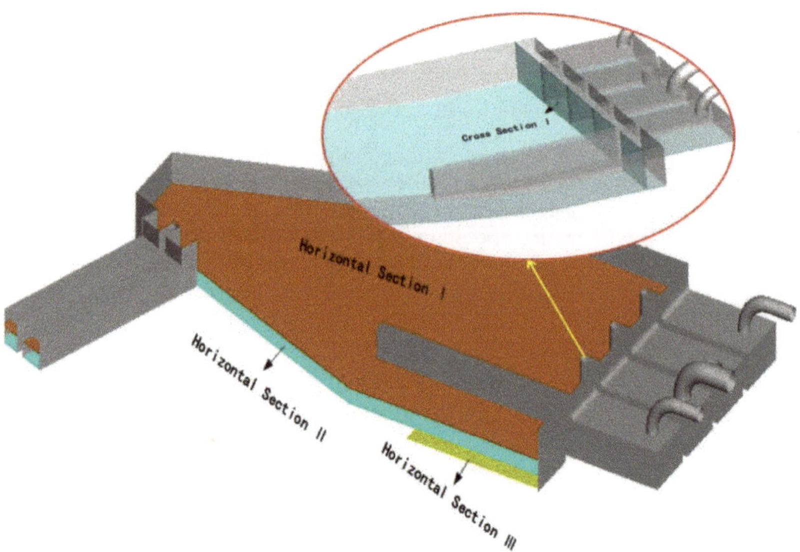

Figure 4. Typical sections.

The leftmost section of the forebay at length direction is Y = 0. Cross-section I (Y = 49 m) is selected at the inlet passage of every pump to calculate the uniformity of axial flow velocity and weighted average angle of flow velocity. The uniformity of axial flow velocity (UAFV) and weighted average angle of flow velocity (WAAFV) on the cross-section I can quantitatively indicate the inlet condition of pumps.

The uniformity of axial flow velocity, V_{au}, is shown in Equation (2).

$$V_{au} = \left(1 - \sqrt{\sum_{i=1}^{n} (V_{ai}/V_a - 1)^2 / n}\right) \times 100\% \tag{2}$$

In the formula, V_{ai} is the axial velocity of each mesh, m/s; V_a is the average axial velocity of the section, m/s; and n is the number of mesh.

The weighted average angle of flow velocity, θ, is shown in Equation (3); θ = 90° indicates that the inlet condition of the pump is very good.

$$\theta = \frac{\sum V_{ai}\left(90^\circ - arctan\frac{V_{ti}}{V_{ai}}\right)}{\sum V_{ai}} \tag{3}$$

In the formula, V_{ti} is the transverse velocity of each mesh, m/s.

4. Design of Diversion Piers Based on Orthogonal Test Method

4.1. Dimension Design of Diversion Piers

As shown in Figure 5, there are three diversion piers: 1 (outer), 2 (middle), and 3 (inner) in consideration of the lateral intake forebay. The diversion piers are designed by arc section and the straight-line segment at the head and tail. Diversion piers 2 and 3 are designed with a straight segment at the head L1 = 4.5 m with the arc segment l1 = 6.7 m in diversion pier 1 as the reference length. The length of the straight-line segment at the head and the length of the arc segment of diversion piers 2 and 3 were determined by relative length which is the ratio between the lengths of the two straight-line segments or two arc segments, L1/L2 = l1/l2. C is the length of the straight-line segment at the tail of diversion pier 3, and the length of the straight-line segment at the tail of diversion piers 1 and 2 may be determined to keep the ends of three diversion piers in the same cross-section of the forebay. Relative height is h = H1/H2, where H1 is the height of the diversion pier and H2 is the water depth of the forebay. B is the width of the diversion pier which is smaller than 0.9 m.

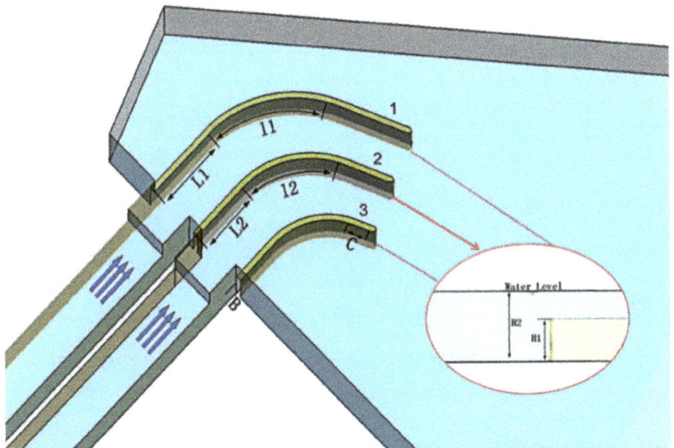

Figure 5. Dimension design of the diversion piers.

4.2. Factors and Results Based on Orthogonal Test Method

The relative length, the width of the diversion pier, the length of the straight-line segment at the tail of the diversion pier, and the relative height are the design parameters for the design of the diversion pier. The uniformity of axial flow velocity and the co-angle of the weighted average angle of flow velocity (CWAAFV) were taken as the evaluation indexes for the orthogonal test results, and the orthogonal test table $L_9(3^4)$ was selected according to the specified factors and the appropriate amount of the level. The level of each factor and the orthogonal test results are shown in Tables 1 and 2. A, B, C, and D represent the relative length, the width of the diversion pier, the length of the straight-line segment at the tail of the diversion pier, and the relative height, respectively.

Table 1. Three levels with four factors.

Level	Factors			
	A	**B (m)**	**C (m)**	**D**
1	1.2	0.3	1	0.6
2	1.3	0.4	1.2	0.8
3	1.4	0.5	1.4	1

Table 2. Orthogonal test results.

Test Number	Factors				Results of Test	
	A	B	C	D	UAFV (%)	CWAAFV (°)
1	1	1	1	1	85.665	7.398
2	1	2	2	2	88.528	22.473
3	1	3	3	3	79.698	37.819
4	2	1	2	3	76.714	33.429
5	2	2	3	1	82.426	12.775
6	2	3	1	2	80.727	14.701
7	3	1	3	2	73.546	25.101
8	3	2	1	3	82.402	31.221
9	3	3	2	1	77.980	13.649

4.3. Analysis of Orthogonal Test Table of the Diversion Piers

Ki (i = 1,2,3) of factor I (I = A,B,C,D) for UAFV or CWAAFV in Table 3 is the sum of three UAFV or CWAAFV values for level i and factor I in Table 2, and then ki = Ki/3, $R = \{\max(k1, k2, k3) - \min(k1, k2, k3)\}$. The R value can reflect the strength of the factor on the result, that is, the larger the R value, the greater the influence of the factor on the result. It can be seen from Table 3 that the influence of UAFV is: A > B > C > D, and the order of the design factors of the diversion piers is: the relative length, the width of the diversion pier, the length of the straight-line segment at the tail and the relative height of the diversion pier. It can also be seen from Table 3 that the influence of CWAAFV is: D > C > A > B, and the order of the design factors of the diversion piers is the relative height of the diversion pier, the length of the straight-line segment at the tail, the relative length, and the width of the diversion pier.

Table 3. Range analysis table.

Range Analysis	UAFV				CWAAFV			
	A	B	C	D	A	B	C	D
K1	253.890	235.925	248.794	246.071	67.690	65.927	53.319	33.821
K2	239.867	253.356	243.222	242.801	60.904	66.469	69.551	62.274
K3	233.928	238.405	235.670	238.814	69.970	66.168	75.695	102.469
k1	84.630	78.642	82.931	82.024	22.563	21.976	17.773	11.274
k2	79.956	84.452	81.074	80.934	20.301	22.156	23.184	20.758
k3	77.976	79.468	78.557	79.605	23.323	22.056	25.232	34.156
R	6.654	5.810	4.375	2.419	3.022	0.181	7.458	22.883

The trend diagram of UAFV is shown in Figure 6. It can be seen from Figure 5 that the relative length of the diversion pier and the length of the straight-line segment at the tail and the relative height of the diversion pier increases, and the UAFV value decreases, but when the width of the diversion pier (0.3 m, 0.4 m, or 0.5 m) is 0.4 m, the UAFV is maximized. So, the test scheme when UAFV is optimal is: A1B2C1D1.

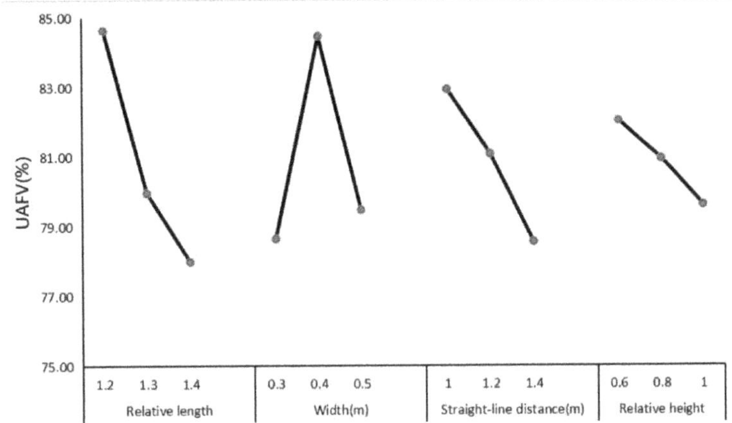

Figure 6. The trend of UAFV with three levels for each factor.

The trend diagram of CWAAFV is shown in Figure 7. When the relative length reaches 1.3 or the width of the diversion pier is 0.3 m or the length of the straight-line segment at the tail is 1.0 m or the relative height of the diversion pier is 0.6 m, CWAAFV values are minimized. So, the optimal scheme for CWAAFV is: A2B1C1D1.

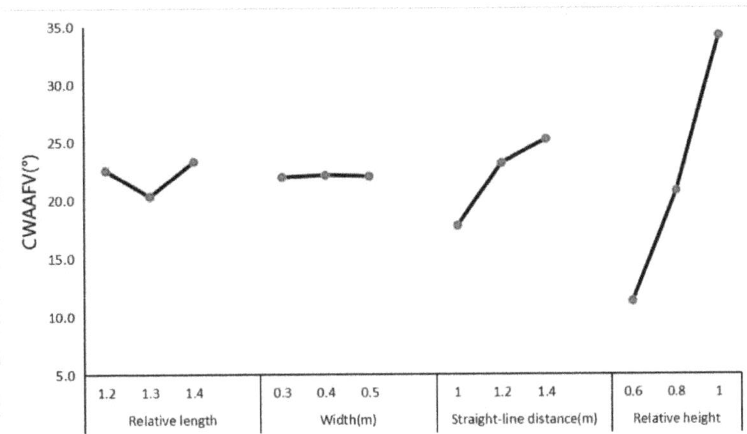

Figure 7. The trend of CWAAFV with three levels for each factor.

4.4. Analysis of Optimal Result

For UAFV, the selection scheme is A1B2C1D1, and for CWAAFV, the selection scheme is A2B1C1D1. The comprehensive balance method is adopted, that is, according to the influence degree of each factor on the evaluation index, the advantages and disadvantages of each factor level for the evaluation index are calculated, and the most favorable scheme is selected. The optimal scheme, A1B2C1D1, of diversion pier design is obtained by calculating with the comprehensive balance method. Based on the numerical simulation, the flow pattern for the optimal scheme is shown in Figure 8. For the No.1 pump, the UAFV was 80.26% and the WAAFV was 77.68°. For the No.2 pump, the UAFV was 98.74% and the WAAFV was 87.84°. For the No.4 pump, the UAFV was 93.41% and the WAAFV was 77.28°.

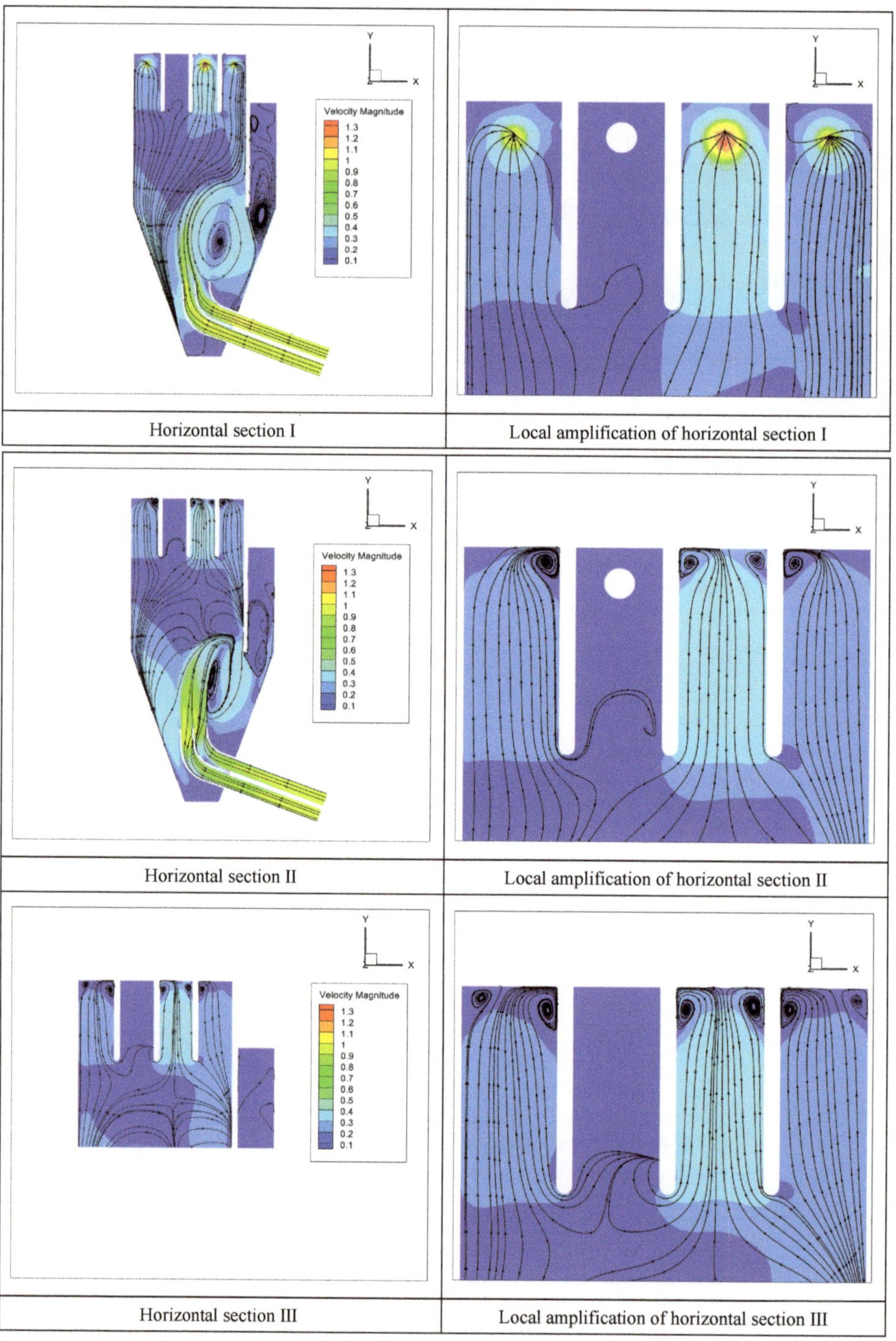

Figure 8. Flow pattern for the optimal scheme of the diversion piers for operating pumps 1, 2, and 4.

Figure 9 is a vector diagram of the flow velocity in the domain. With the increase in the trip, the water flow velocity gradually decreases, the backflow at the outlet of the diversion piers gradually disappears, the flow velocity distribution is gradually uniform, and the direction of the flow velocity at the cross-section I is basically parallel to the axial direction, showing that the diversion pier has a good rectification effect.

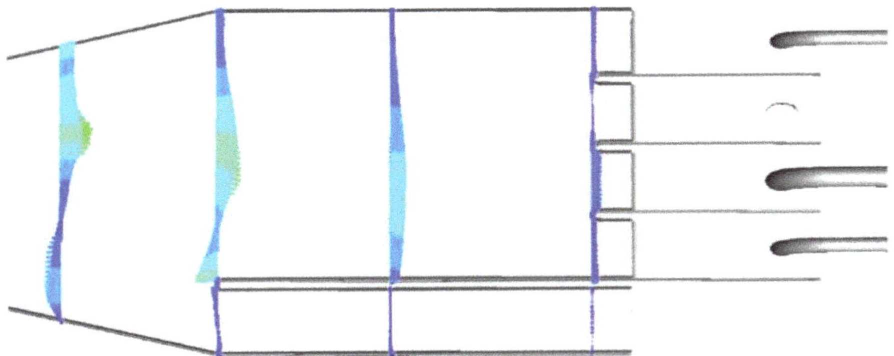

Figure 9. The velocity vector of the approach diagram.

5. CFD Numerical Test Analysis of Pumps 1, 3, and 4

In order to test the rectification effect of the optimal scheme designed in the orthogonal test of the diversion piers under the pump operation scheme of pumps 1, 3, and 4, the flow pattern in the forebay of the pumping station is calculated by the same CFD method, and the streamline diagram of each typical section is shown in Figure 10.

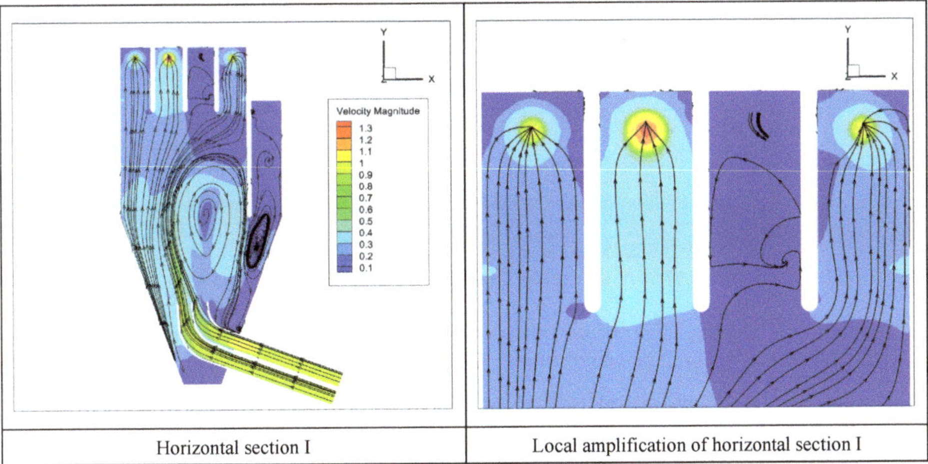

Figure 10. *Cont.*

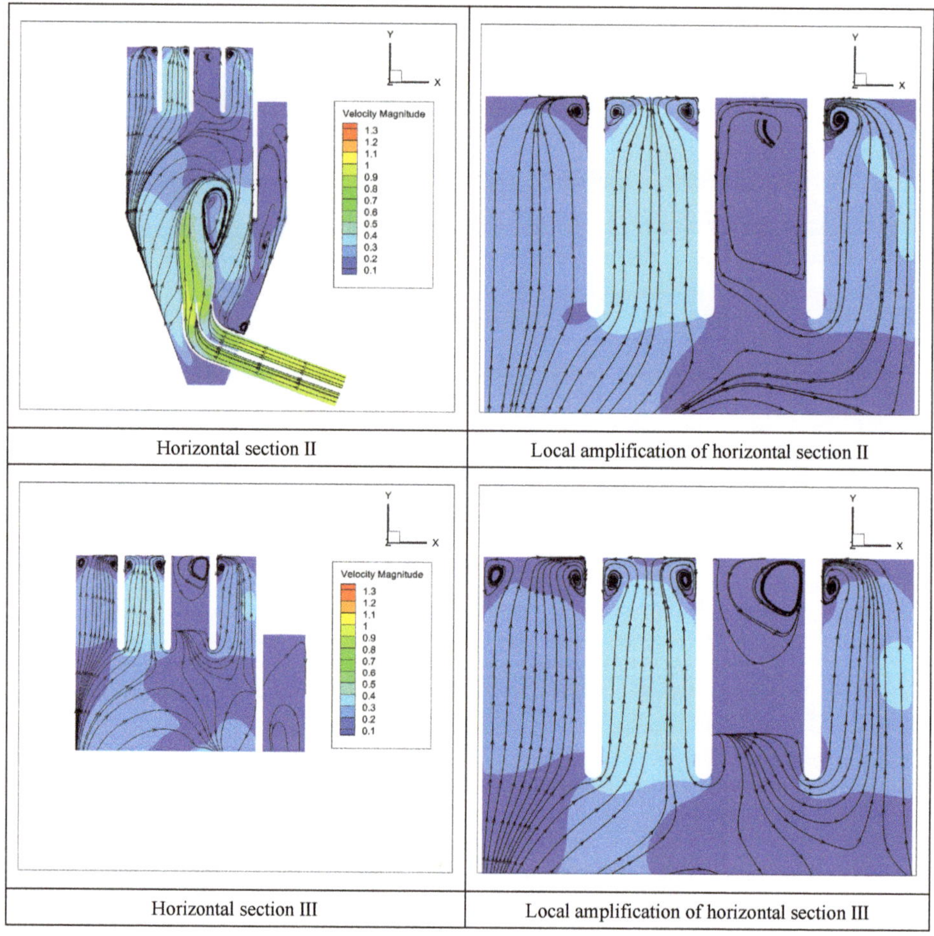

Figure 10. Flow pattern for the optimal scheme of the diversion piers for operating pumps 1, 3, and 4.

By the calculation, the UAFV of the No.1, No.3, and No.4 pumps were 92.65%, 94.54%, and 75.81%, respectively. The WAAFV of pumps 1, 3, and 4 were 72.66°, 85.14°, and 78.21°, respectively. Therefore, it can be proved that the optimal scheme of the diversion piers designed by the orthogonal test method is more suitable for different pump startup modes under lateral water intake.

6. Conclusions

The CFD numerical simulation approach was used to study the flow pattern in the forebay and to improve the design parameters of the diversion pier of a pumping station. The forebay of a lateral intake pumping station with asymmetrical working pumps was subjected to flow analysis using a realizable k-ε Turbulent model, the SIMPLEC method, Pressure Inlet boundary condition, where the pressure between the top surface and the bottom surface is linear with the height of the inlet section, and Mass Flow Outlet boundary condition. Based on the flow parameters, the diversion pier was chosen as the optimization strategy. Nine groups of the numerical simulation on various layouts of diversion piers were examined using the orthogonal test method, with the uniform axial flow velocity and weighted average angle of the flow velocity of each pump's inlet cross-section. This provided a reference for related pumping station engineering. The following conclusions were drawn:

(1) The parameters of diversion piers of the orthogonal test design having four factors and three levels are different, and the uniformity of axial flow velocity and the weighted average angle of the flow velocity of the inlet passages for the pumps are also different. With a comprehensive balance method, the optimal combination of diversion pier parameters can be selected as A1B2C1D1. The uniformity of axial flow velocity and the weighted average angle of flow velocity after selecting the optimal combination are greatly improved.

(2) After selecting the optimal scheme through the orthogonal test design of the diversion piers, the results showed that a small range of reflux phenomena appears in the outlet position of the diversion piers. However, with the increase in the water flow approach, the backflow area gradually disappears, and the water flow is relatively straight. Because the flow of the No.2 pump unit is greater than that of the No.1 pump unit and No.4 pump unit, the velocity distribution of the cross-section of the forebay shows a trend of middle-high and low on both sides.

(3) Under the pump operation scheme of pumps 1, 3, and 4, the rectification effect of the optimal scheme designed in the orthogonal test of the diversion piers under the pump operation scheme of pumps 1, 2, and 4 has been well tested. It can be proved that the size of the diversion pier designed by the orthogonal test method has a high application value in the study of the lateral intake pumping station.

RANS with a realizable k-ε turbulent model was used in this study. The LES turbulent model will be a good approach for higher accuracy and fine analysis of the vortex field. The diversion piers in the paper combined with the sill is likely obtain a better rectification effect, according to the relevant references. Sediment deposition in the front of the sill, however, needs to be taken into consideration; it also necessitates further investigation.

Author Contributions: Conceptualization, C.Z. and H.Y.; methodology, C.Z.; software, C.Z.; validation, H.Y., Y.Y., and M.T.J.; formal analysis, C.Z.; investigation, C.Z.; resources, Y.Y.; data curation, H.Y.; writing—original draft preparation, M.T.J.; writing—review and editing, Y.Y.; visualization, C.Z.; supervision, Y.Y.; project administration, Y.Y.; funding acquisition, Y.Y. All authors have read and agreed to the published version of the manuscript.

Funding: The research was supported by the National Key Research and Development Program (2019YFC0409000).

Institutional Review Board Statement: Not applicable.

Informed Consent Statement: Not applicable.

Conflicts of Interest: The authors declare no conflict of interest.

References

1. Nasr, A.; Yang, F.; Zhang, Y.; Wang, T.; Hassan, M. Analysis of the Flow Pattern and Flow Rectification Measures of the Side-Intake Forebay in a Multi-Unit Pumping Station. *Water* **2021**, *13*, 2025. [CrossRef]
2. Liu, C.; Han, X.; Zhou, J.; Jin, Y.; Cheng, L. Numerical simulation of turbulent flow in forebay with side-intake of pumping station. *J. Drain. Irrig. Mach. Eng.* **2009**, *27*, 281–286.
3. Cheng, B.; Yu, Y. CFD Simulation and Optimization for Lateral Diversion and Intake Pumping Stations. *Procedia Eng.* **2012**, *28*, 122–127. [CrossRef]
4. Ansar, M.; Nakato, T.; Constantinescu, G. Numerical simulations of inviscid three-dimensional flows at single- and dual-pump intakes. *J. Hydraul. Res.* **2002**, *40*, 461–470. [CrossRef]
5. Rajendran, V.P.; Constantinescu, S.G.; Patel, V.C. Experimental Validation of Numerical Model of Flow in Pump-Intake Bays. *J. Hydraul. Eng.* **1999**, *125*, 1119–1125. [CrossRef]
6. Rajendran, V.P.; Patel, V.C. Measurement of Vortices in Model Pump-Intake Bay by PIV. *J. Hydraul. Eng.* **2000**, *126*, 322–334. [CrossRef]
7. Teaima, I.R.; El-Gamal, T. Improving Flow Performance of Irrigation Pump Station Intake. *J. Appl. Water Eng. Res.* **2017**, *5*, 9–21. [CrossRef]
8. Zhan, J.-M.; Wang, B.-C.; Yu, L.-H.; Li, Y.-S.; Tang, L. Numerical Investigation of Flow Patterns in Different Pump Intake Systems. *J. Hydrodynam. B* **2012**, *24*, 873–882. [CrossRef]

9. Kadam, P.; Chavan, D. CFD analysis of flow in pump sump to check suitability for better performance of pump. *Int. J. Mech. Eng. Robot.* **2013**, *1*, 56–65.
10. Liu, C.; Zhou, J.; Cheng, L. The experimental study and numerical simulation of turbulent flow in pumping forebay. *ASME Power Conf.* **2009**, *43505*, 171–176.
11. Liu, C.; Zhou, J.; Cheng, L.; Jin, Y.; Han, X. Study on Improving the Flow in Forebay of the Pumping Station. In Proceedings of the ASME 2010 3rd Joint US-European Fluids Engineering Summer Meeting: Volume 1, Symposia—Parts A, B, and C; ASMEDC, Montreal, QC, Canada, 1–5 August 2010.
12. Feng, X. Flow analysis of bottom sill rectification and back sill of pump station forebay. *Jiangsu Water Resour.* **1998**, *26*, 31–33, 38.
13. Li, J.; Cao, Y.; Gao, C. Numerical Simulation of Flow Patterns in the Forebay and Suction Sump of Tianshan Pumping Station. *Water Pract. Technol.* **2014**, *9*, 519–525. [CrossRef]
14. Xu, B.; Zhang, C.; Li, Z.; Gao, C.; Bi, C. Study on the influence of geometric parameters of diversion pier on navigable flow conditions of gate station joint hub based on CFD. *J. Irrig. Drain.* **2019**, *38*, 115–122.
15. Xia, C.; Cheng, L.; Jiao, W.; Zhang, D. Numerical simulation on rectification measure of inverted T-shaped sill at forebay of pump station. *J. South-to-North Water Transf. Water Sci. Technol.* **2018**, *16*, 146–150.
16. Luo, C.; Cheng, L.; Liu, C. Numerical simulation of mechanism for sill rectifying flow in pumping station intake. *J. Drain. Irrig. Mach. Eng.* **2014**, *32*, 393–398.
17. Yu, Y.; Xu, H.; Cheng, Y. CFD numerical simulation on modification of flow pattern with flow deflector at fore-bay of pumping station. *J. Water Resour. Hydropower Eng.* **2006**, *37*, 41–43.
18. Luo, C.; Qian, J.; Liu, C.; Chen, F.; Xu, J.; Zhou, J. Numerical simulation and test verification on diversion pier rectifying flow in forebay of pumping station for asymmetric combined sluice-pump station project. *Trans. Chin. Soc. Agric. Eng.* **2015**, *31*, 100–108.
19. Wang, W.; Shi, W.; Jiang, X.; Feng, Q.; Lu, W.; Zhang, D. Optimization design of multistage centrifugal pump impeller by orthogonal experiment and CFD. *Appl. Energy* **2016**, *34*, 191–197.
20. Zhou, J.; Zhao, M.; Wang, C.; Gao, Z. Optimal Design of Diversion Piers of Lateral Intake Pumping Station Based on Orthogonal Test. *Shock Vib.* **2021**, *2021*, 1–9. [CrossRef]
21. Xu, B.; Liu, J.; Lu, W. Optimization Design of Y-Shaped Settling Diversion Wall Based on Orthogonal Test. *Machines* **2022**, *10*, 91. [CrossRef]
22. Zhou, J.; Zhong, Z.; Liang, J.; Shi, X. Three-dimensional Numerical Simulation of Side-intake Forebay of Pumping Station. *J. Irrig. Drain.* **2015**, *34*, 52–55.
23. Yang, F.; Zhang, Y.; Liu, C.; Wang, T.; Jiang, D.; Jin, Y. Numerical and Experimental Investigations of Flow Pattern and Anti-Vortex Measures of Forebay in a Multi-Unit Pumping Station. *Water* **2021**, *13*, 935. [CrossRef]
24. Can, L.; Chao, L. Numerical simulation and improvement of side-intake characteristics of multi-unit pumping station. *J. Hydroelectr. Eng.* **2015**, *34*, 207–214.
25. Caishui, H.O.U. Three-Dimensional Numerical Analysis of Flow Pattern in Pressure Forebay of Hydropower Station. *Procedia Eng.* **2012**, *28*, 128–135. [CrossRef]
26. Lu, Z.; Xiao, R.; Tao, R.; Li, P.; Liu, W. Influence of guide vane profile on the flow energy dissipation in a reversible pump-turbine at pump mode. *J. Energy Storage* **2022**, *49*, 104161. [CrossRef]
27. Wang, F. *The Analysis of Computational Fluid Dynamics-CFD Software Theory and Application*; Tsinghua University Press: Beijing, China, 2004.

Article

Analysis of Hydraulic Performance and Flow Characteristics of Inlet and Outlet Channels of Integrated Pump Gate

Chuanliu Xie [1,*], Weipeng Xuan [1], Andong Feng [1] and Fei Sun [2]

[1] College of Engineering, Anhui Agricultural University, Hefei 230036, China
[2] Suqian Branch of Jiangsu Water Source Company of South to North Water Diversion, Suqian 223801, China
* Correspondence: xcltg@ahau.edu.cn

Abstract: The integrated pump gate structure can improve the shortcomings of traditional asymmetric pumping stations with large floor space, but its internal flow mechanism is not clear, which affects its efficient, stable, and safe operation. In order to reveal its internal fluid flow characteristics, numerical simulations based on the N-S equation with the SST k-ω turbulence model are used in this paper, and experimental validation is carried out. The test results yielded an efficiency of 60.50% near the design flow condition, corresponding to a flow rate of 11.5 L/s, a head of 2.7569 m, a hydraulic loss of 0.064 m in the inlet channel, and a hydraulic loss of 1.337 m in the outlet channel. The integrated pump gate has a uniform inlet water flow pattern, less undesirable flow pattern, and a large backflow vortex in the outlet water. This paper reveals the internal flow characteristics of its integrated pump gate inlet and outlet water, and the research results can provide some reference for the design, theoretical analysis, and application of similar integrated pump gates.

Keywords: integrated pump gate; inlet channel; outlet channel; flow pattern; hydraulic performance

Citation: Xie, C.; Xuan, W.; Feng, A.; Sun, F. Analysis of Hydraulic Performance and Flow Characteristics of Inlet and Outlet Channels of Integrated Pump Gate. *Water* **2022**, *14*, 2747. https://doi.org/10.3390/w14172747

Academic Editors: Ran Tao, Changliang Ye and Xijie Song

Received: 28 July 2022
Accepted: 30 August 2022
Published: 2 September 2022

Publisher's Note: MDPI stays neutral with regard to jurisdictional claims in published maps and institutional affiliations.

Copyright: © 2022 by the authors. Licensee MDPI, Basel, Switzerland. This article is an open access article distributed under the terms and conditions of the Creative Commons Attribution (CC BY) license (https://creativecommons.org/licenses/by/4.0/).

1. Introduction

In recent years, the frequent occurrence of extreme rainfall weather has posed a serious challenge to the flood control and drainage capacity of cities and towns. In the original pumping station construction, the asymmetric arrangement form is usually used, that is, the combination of sluice gate and pumping station, the gate is set in the middle of the river cross-section direction, and the pump room is set next to the sluice gate. Usually, the area of the water barrier sluice is small, and the pump chambers on both sides cannot pass through the water when no flooding work is needed, resulting in low water exchange efficiency inside and outside the river. At the same time, the construction of a large area of land, long construction period, high economic costs are the traditional pumping station disadvantages. Therefore, the integrated pump and gate device came into being. The integrated pump and gate device is a pumping station that combines a pumping station and a water barrier sluice into a whole arrangement, in which the pump sluice can be used as a water barrier structure instead of a traditional sluice gate, and can provide support for the pump. The pump is arranged on the gate, so there is no need to build a fixed pump room, so that the gate and the pump room are combined into a whole device, which greatly improves the flooding capacity of the river, significantly increases the vitality of the water body in the river channel, and has the advantages of occupying a small area of land compared to traditional pumping stations, short project cycle, lifting and handling, and automatic control, etc., which can also significantly reduce the construction cost, land cost, and alteration cost, and so on, while improving the efficiency of water exchange.

The integrated pump gate consists of sluice gate, gate pump, clapper gate, water stop structure, opening and closing mechanism, etc. Relevant scholars have analyzed its application characteristics and technical advantages, and compared it with traditional pump gate in terms of floor space, circulation, economy, construction cycle, etc. Considering that integrated pump gates play a significant role in controlling drainage within urban areas and managing black smelly water bodies [1]. Some scholars have discussed their design points and concluded that in urban and rural river management, integrated pump gates will become the mainstream and where the trend of development lies [2]. Some scholars have explored their structures, such as the design of horizontal axial submersible pumps suitable for pump gates [3], to investigate the effects of pump form, installation number, overhang height, pump spacing, and flapper angle on the performance of integrated pump gates [4]. Some other scholars have conducted numerical calculations and model experimental studies on pump gates to derive the stress conditions, flow conditions, and gate vibration characteristics of vertical surface-hole integrated pump gates and horizontal surface-hole integrated pump gates under different operating conditions [5,6]. There is always a negative pressure zone directly below the bottom edge of the flat bottom gate [7], suggesting the two-phase flow and pump gate characteristics in the pump gate pool [8], testing the energy characteristics of the ecological gate pump during bi-directional operation, and resulting in a error of ±2% with the numerical simulation, which verified the reasonableness of the numerical simulation, and finally predicted the optimal installation angle of the vane by numerical simulation [9]. These research results provide basic data for subsequent pump gates selection and other aspects, and also provide reference for subsequent pump gates design and research.

At present, the integrated pump gate has formed a certain scale of industrial system and user demand, but the related research is less, and the flow pattern of the front pool and inlet and outlet water channels is not clear, which limits the further development of the integrated pump gate technology, so it is necessary to carry out further related research, and the research results of this paper have certain theoretical significance and engineering value.

2. Numerical Simulation Model and Method

2.1. Pump Gate Modeling

In accordance with a project pump station design, impeller wood mold diagram, pump gate structure diagram, and other applications of SolidWorks software were used to establish an integrated pump gate model, imported into ANSYS Workbench software [10] in the geometry module for adjustment and boundary condition naming; where model impeller diameter D = 60 mm, rotation rate n = 6692 r/min, the gate length is 188 mm, width is 60 mm and height is 299 mm, integrated pump gate design flow rate Q_d = 11.5 L/s, design head H = 2.7569 m. Three-dimensional geometric model of integrated pump gate as shown in Figure 1.

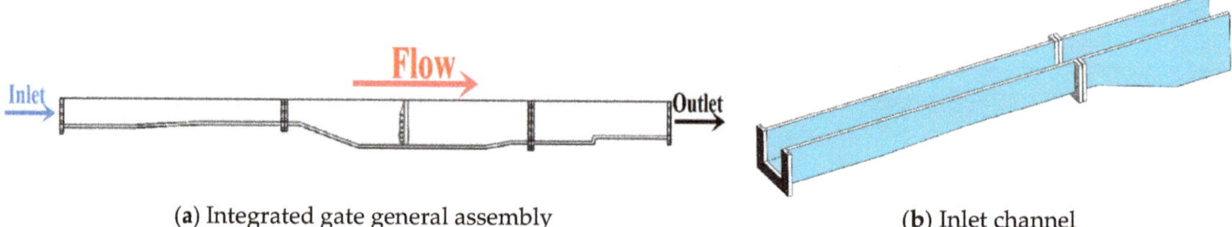

(**a**) Integrated gate general assembly

(**b**) Inlet channel

Figure 1. *Cont.*

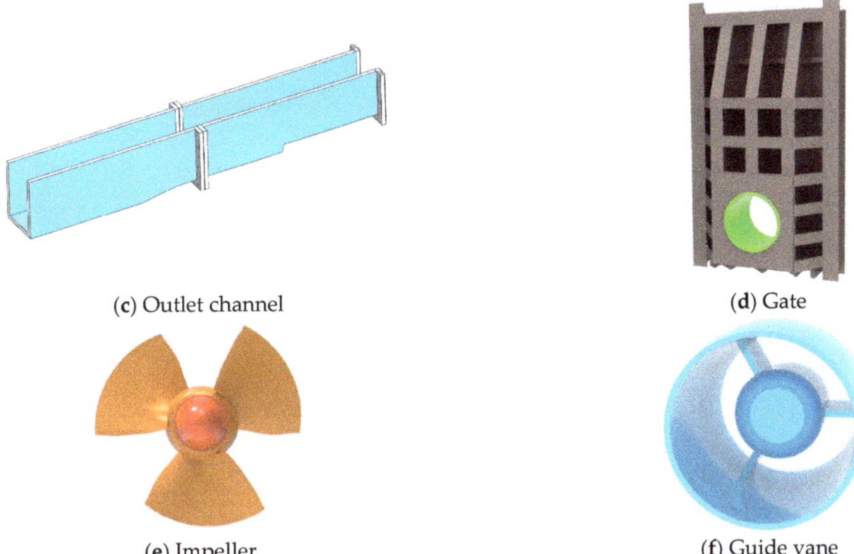

Figure 1. Three-dimensional geometric model of integrated pump gate. (**a**) Integrated gate general assembly; (**b**) Inlet channel; (**c**) Outlet channel; (**d**) Gate; (**e**) Impeller; (**f**) Guide vane.

2.2. Mesh Division

According to the three-dimensional structure of the integrated pump gate, the water overflow part of it is extracted, and the numerical simulation area of the integrated pump gate includes: open inlet channel, gate, impeller, guide vanes, and open outlet channel. In the calculation of this paper, the integrated pump gate is a table-hole type, and in order to better retain the characteristics of the pump gate surface, the model is divided by unstructured mesh, and the number of geometric model meshes is changed under the design flow condition (Q_d = 11.5 L/s) for mesh-independent analysis. After the grid number reaches 3.03 million, the efficiency of the pump gate basically does not change with the increase of the grid number [11], and this paper determines how to use the grid number of 3.03 million for the subsequent numerical simulation work. Mesh division of each calculation component as shown in Figure 2. The grid-independence results are shown in Figure 3.

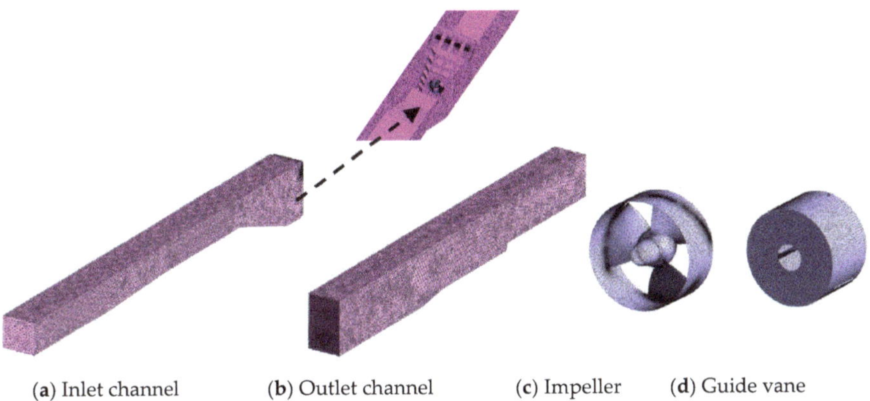

Figure 2. Mesh division of each calculation component.

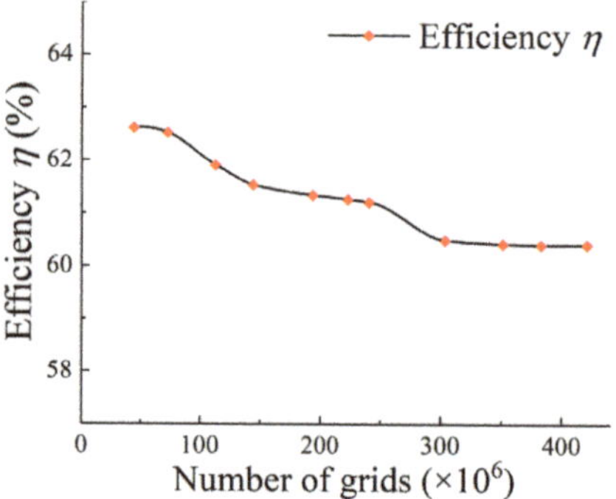

Figure 3. Grid-independent analysis.

2.3. Control Equations, Boundary Conditions, and Calculation Methods

In the numerical simulation of the integrated pump gate, the fluid is treated as a three-dimensional incompressible viscous turbulent flow, without considering heat exchange. Boundary condition settings as shown in Table 1.

Table 1. Boundary condition settings.

Boundary Conditions	Parameter Setting
Inlet	Mass Flow Rate
Outlet	Average Static Pressure
Free liquid level	Symmetry
Dynamic and static interface	Frozen Rotor
Static interfaces	None

In this paper, the SST k-ω [12] turbulence model is used to calculate the flow characteristics inside the integrated pump gate, which combines the advantages of the standard k-ε model [13] and the standard k-ω model [14] and captures the flow in the boundary layer better by using automatic functions in the boundary layer, while the finite element-based finite volume method is used to solve the problem.

In order to evaluate the hydraulic performance of the integrated pump gate after numerical simulation, the hydraulic losses of the inlet and outlet channels, the uniformity of the axial velocity distribution in the characteristic section, the weighted average angle of the velocity in the characteristic section and the energy characteristics of the pump gate were introduced as evaluation bases, and the performance characteristics of the integrated pump gate were evaluated by five quantitative indicators.

In this paper, the total energy difference between the inlet channel of the integrated pump gate and the outlet of the outlet channel is defined as the head of the device and is expressed by the following equation [15]:

$$H_{net} = \left(\frac{\int\limits_{s2} P_2 u_t ds}{\rho Q g} + H_2 + \frac{\int\limits_{s2} u_2^2 u_{t2} ds}{2Qg} \right) - \left(\frac{\int\limits_{s1} P_1 u_t ds}{\rho Q g} + H_1 + \frac{\int\limits_{s1} u_1^2 u_{t1} ds}{2Qg} \right) \quad (1)$$

where the first term on the right side of the equation is the total pressure at the outlet of the outlet channel and the second term is the total pressure at the inlet of the inlet channel.

Where Q is the flow rate (m^3/s), H_1, H_2 are the inlet and outlet section elevations of the integrated pump gate (m), s_1, s_2 for the integrated pump gate inlet and outlet section, u_1, u_2 for the integrated pump gate inlet and outlet water channel section flow rate at each point (m/s), u_{t1}, u_{t2} are the normal components of flow velocity (m/s) at each point of the inlet and outlet channel sections of the integrated pump gate, P_1, P_2 are the static pressure

(Pa) at each point of the inlet and outlet sections of the integrated pump gate, g is the acceleration of gravity (m/s^2).

The efficiency of the integrated pump gate is calculated as [16,17]:

$$\eta = \frac{\rho g Q H_{net}}{T_p w} \quad (2)$$

where T_p is the torque (N-m), ω is the rotational angular speed of the impeller. The hydraulic loss h_f is calculated as [18,19]:

$$h_f = E_1 - E_2 = (\frac{P_1}{\rho g} - \frac{P_2}{\rho g}) + (Z_1 - Z_2) + (\frac{{u_1}^2}{2g} - \frac{{u_2}^2}{2g}) \quad (3)$$

Among them:

$$E_1 = \frac{P_1}{\rho g} + Z_1 + \frac{{u_1}^2}{2g};\ E_2 = \frac{P_2}{\rho g} + Z_2 + \frac{{u_2}^2}{2g}$$

where E_1, E_2 are the total energy at the inlet and outlet of the open flow channel, P_1, P_2 are the static pressure at the inlet and outlet of the open flow channel (Pa), Z_1, Z_2 are the height of the open flow channel inlet and outlet (m), u_1, u_2 are the open flow channel inlet and outlet velocity (m/s).

The uniformity of flow velocity distribution is calculated as [20]:

$$V_u = \left\{ 1 - \frac{1}{\overline{v_a}} \sqrt{\left[\sum_{i=1}^{n} (v_{ai} - \overline{v_a})^2 \right] / n} \right\} \times 100\% \quad (4)$$

where, V_u is the uniformity of axial flow velocity distribution in the characteristic section (%), v_{ai} is the axial velocity of each calculation unit (m/s), n is the number of calculation units. The velocity weighted average angle is calculated as [21]:

$$\overline{\theta} = \frac{\sum u_{ai} \left[90^\circ - \arctan\left(\frac{u_{ti}}{u_{ai}} \right) \right]}{\sum u_{ai}} \quad (5)$$

where u_{ti} is the horizontal velocity (m/s) of each unit in the characteristic section of the flow channel. u_{ai} is the axial velocity (m/s) of each calculation unit.

3. Numerical Simulation Results and Analysis

3.1. Hydraulic Performance Results and Analysis

The pressure and torque are extracted from the numerical simulation result file, and the pump gate head is calculated according to equation (1), and the pump gate efficiency is calculated according to equation (2) to obtain the energy performance of the integrated pump gate, as shown in Table 2. The energy performance curve of the integrated pump gate is plotted as shown in Figure 4.

Table 2. Numerical simulation of the energy performance of an integrated pump gate.

Q (L/s)	*H* (m)	*H* (%)
8.5 (0.74 Q_d)	3.4884	49.83
9.5 (0.83 Q_d)	3.3126	55.85
10.5 (0.91 Q_d)	3.0206	59.01
11.5 (Q_d)	2.7569	60.50
12.5 (1.09 Q_d)	2.3343	59.73
13.5 (1.17 Q_d)	1.7245	54.87
14.5 (1.26 Q_d)	1.0426	43.58

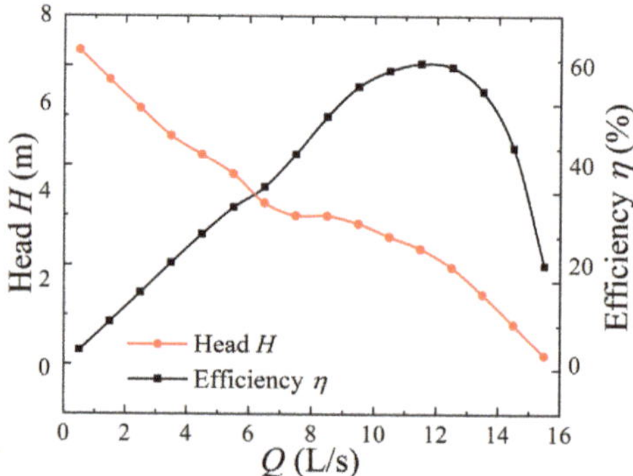

Figure 4. Pump gate energy performance curve.

The calculation results in Table 2 and Figure 4 show that the flow-efficiency curve of the pump gate is approximately quadratic when the inlet water flow is Q = 0.5~15.5 L/s, and the head of the pump gate gradually decreases from 7.24 m to 0.34 m, and the efficiency of the pump gate is 60.50% near the design flow condition, the corresponding flow is 11.5 L/s and the head is 2.7569 m. When the flow rate is 10.5~12.5 L/s (0.91~1.09 Q_d), the pump gate is in the high efficiency zone, and the efficiency of the pump gate is around 59~60%. When the flow rate is 5.5~7.5 L/s (0.48~0.65 Q_d), the pump gate is located near the saddle area, and the operation of the pump gate is not stable at this time, so it is recommended to avoid operating in this flow range.

3.2. Analysis of Internal Flow in Inlet Channel

An axial section was created using the impeller rotation axis as a reference to explore the internal flow characteristics of the inlet flow channel before the pump gate. The schematic diagram of the axial section is shown in Figure 5.

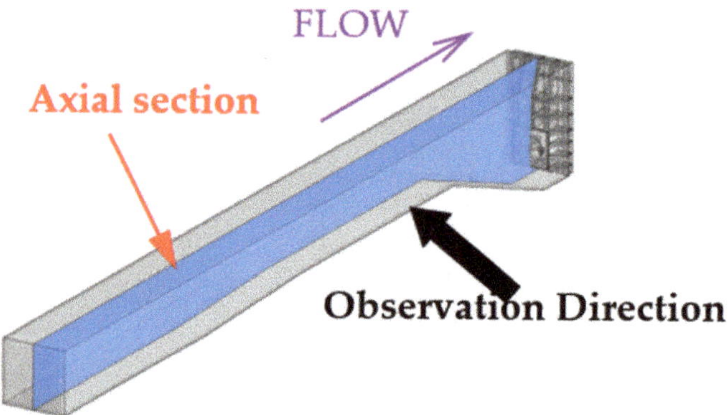

Figure 5. Schematic diagram of the axial section of the open inlet channel.

3.2.1. Inlet Channel Streamline and Axial Flow Velocity Distribution

The streamline and axial flow velocity distribution clouds of the open inlet channel under different flow conditions in the section are drawn, as shown in Figure 6.

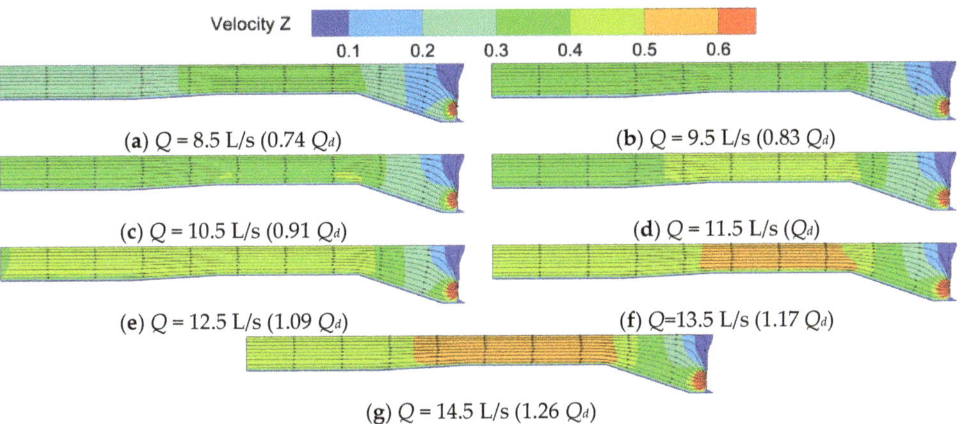

Figure 6. Cloud diagram of inlet channel streamline and axial flow velocity distribution under different working conditions.

Through the analysis of Figure 6, it is concluded that under different flow conditions, the streamline of open inlet channel is smooth, contraction is reasonable, and the streamline is regular, which can provide better water inlet conditions for the impeller. The slope of the open inlet channel is not easy to be too large, as too large will not only lead to aggregation of water flow, but also very easy to form vortex at the bottom, hence should be avoided; the figure can be found in the inlet channel of the slope of the decline is reasonable, no bad flow conditions.

Under each flow condition (Q = 8.5~14.5 L/s), the flow velocity increases in the direction from the inlet of the inlet channel to the inlet of the pump gate; at the top of the pump gate structure, there is a low velocity zone because the water flow receives the constraint of the pump gate; the water flow gathers at the inlet of the pump because of the negative pressure created when the pump rotates, and the flow velocity reaches the maximum value here, which is greater than 0.6 m/s. The flow velocity at the inlet of the pump gate varies in a uniform gradient. Under the design condition (Q = 11.5 L/s), the velocity field is the most uniform in the inlet channel, and under low and high flow conditions, the high and low velocity areas are mixed more obviously.

The local area from the slope of the inlet channel to the inlet of the pump gate is extracted separately as a research object, and the streamline and pressure distribution clouds at the inlet of the pump gate of the open inlet channel under different flow conditions are drawn in the section, as shown in Figure 7.

It can be concluded from Figure 7 that with the increase of flow rate (Q = 8.5~14.5 L/s), the static pressure in the inlet channel is gradually increasing, and under the design condition (Q = 11.5 L/s), the pressure at the bottom of the slope to the pump inlet is approximately the same, and the pressure only starts to decrease in front of the pump gate impeller inlet, under the high flow rate and low flow rate conditions, the pressure gradient near the pump gate inlet is large and the pressure gradually decreases, the pressure is lowest at the inlet of the pump gate.

3.2.2. Hydraulic Loss of Inlet Channel

According to the Formula (3) to calculate the inlet channel hydraulic loss, drawing different flow conditions in the inlet channel hydraulic loss curve, as shown in Figure 8.

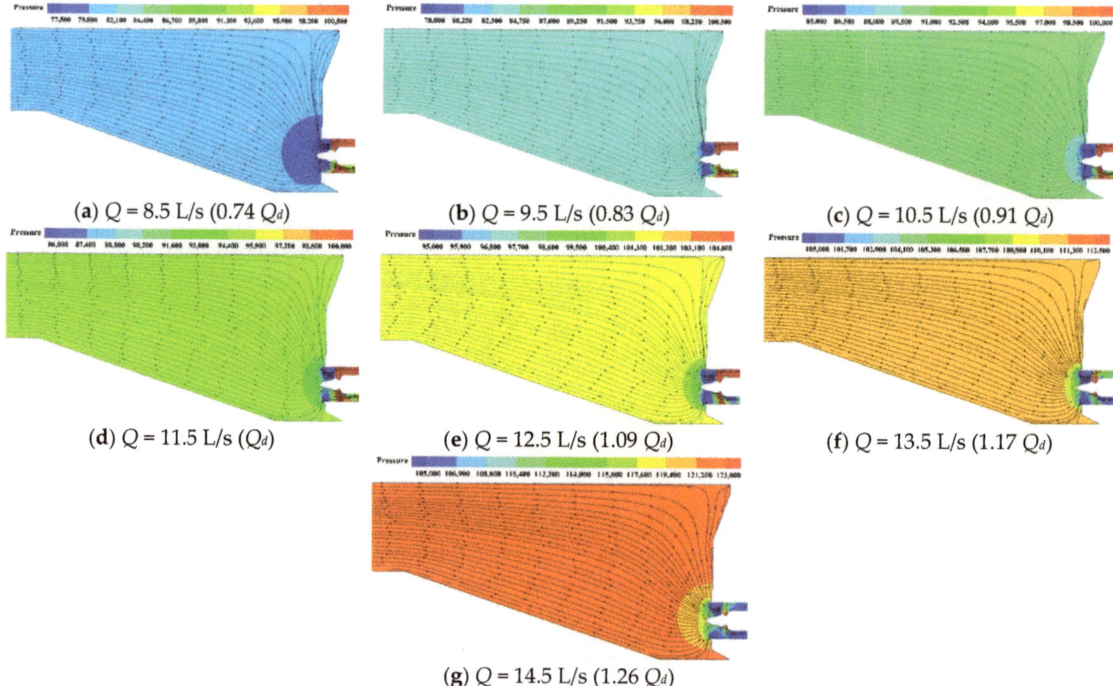

Figure 7. Cloud diagram of streamline and pressure distribution at the inlet of pump gate under different working conditions.

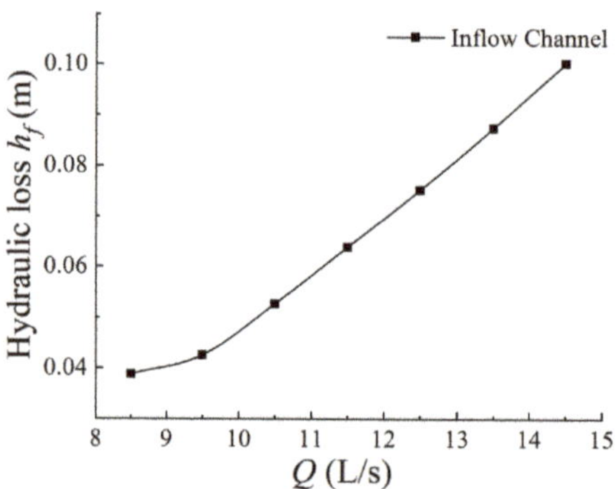

Figure 8. Hydraulic loss curve of inlet channel under different working conditions.

The analysis of Figure 8 shows that the hydraulic loss of the inlet channel h_f is positively correlated with the flow rate Q, which approximately satisfies the quadratic function, and the hydraulic loss is the smallest when the inlet flow rate is 8.5 L/s (0.74 Q_d), which is 0.039 m, and the largest when the flow rate is 14.5 L/s (1.26 Q_d), which is 0.100 m. The larger the flow rate, the larger the hydraulic loss, and in this type of pump station, the hydraulic loss of the inlet and outlet channels is a decisive factor in the efficiency of the pump gate. The calculation results show that the average level of hydraulic loss of the inlet channel is about 6 cm, which is in line with the conventional theory and design.

3.3. Analysis of Internal Flow in Outlet Channel

An axial section is created with the impeller rotation axis to observe the flow characteristics in the outflow channel. The schematic diagram of the axial section is shown in Figure 9.

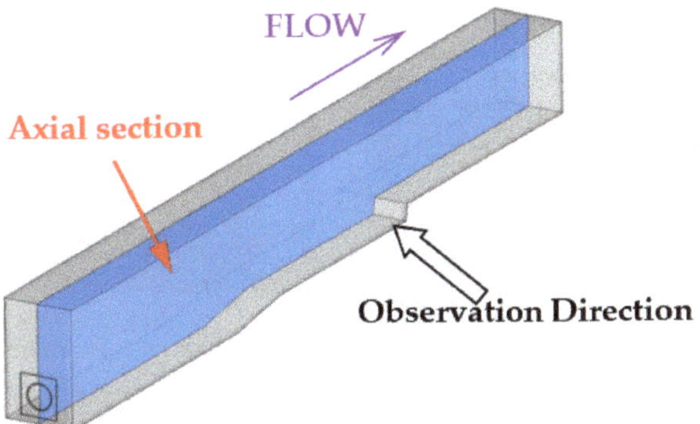

Figure 9. Schematic diagram of axial section of open outlet channel.

3.3.1. Streamline and Axial Velocity Distribution of Outlet Channel

The streamline and axial flow velocity distribution clouds of the open outlet channel under different flow conditions in the section were drawn, as shown in Figure 10.

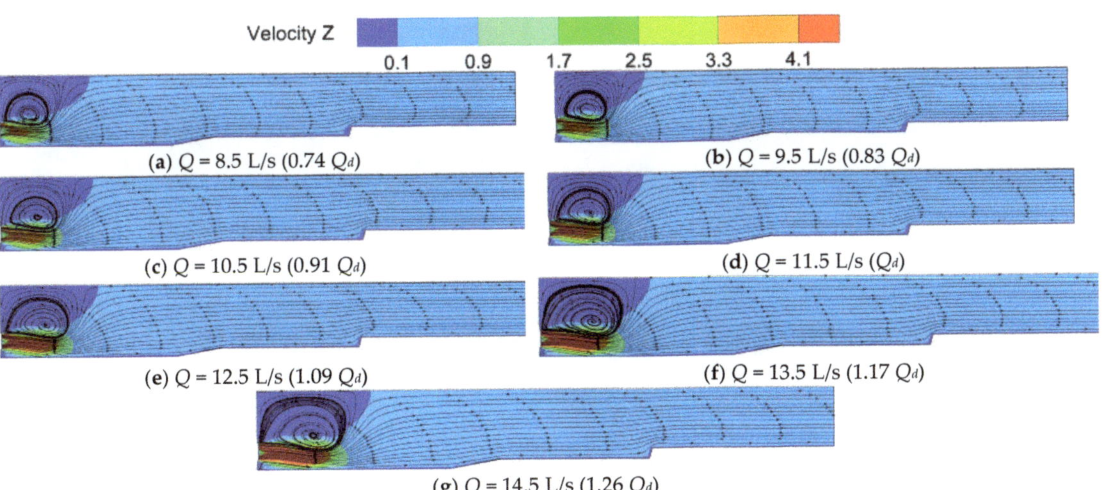

Figure 10. Streamline and axial velocity distribution of the outlet channel under different flow conditions.

It can be found in Figure 10 that the outflow channel is turbulent near the pump gate, and a return vortex parallel to the flow direction is formed at the outlet of the pump gate guide vane, and the size of the vortex increases as the flow rate increases. When the flow rate is 8.5~10.5 L/s (0.74~0.91 Q_d), the axial flow velocity at the outlet of the pump gate is about 3.3 m/s; when the flow rate is 11.5~14.5 L/s (Q_d~1.26 Q_d), the axial flow velocity at the outlet is about 4.1 m/s, and then there is a step transition along the outlet direction, due to the existence of vortex, the axial flow velocity above the pump gate is lower. The axial flow velocity above the outlet is low. The existence of the vortex is mainly due to the open outlet upper part of the stagnant water area, where the flow velocity is low due to the rotation of the impeller, the lower and middle water flow velocity is fast, making the formation parallel to the direction of the water flow back vortex.

The local flow field at the outlet of the pump gate is taken out, and the streamline and pressure distribution clouds at the outlet of the pump gate of the open outflow channel under different flow conditions in the section are drawn, as shown in Figure 11.

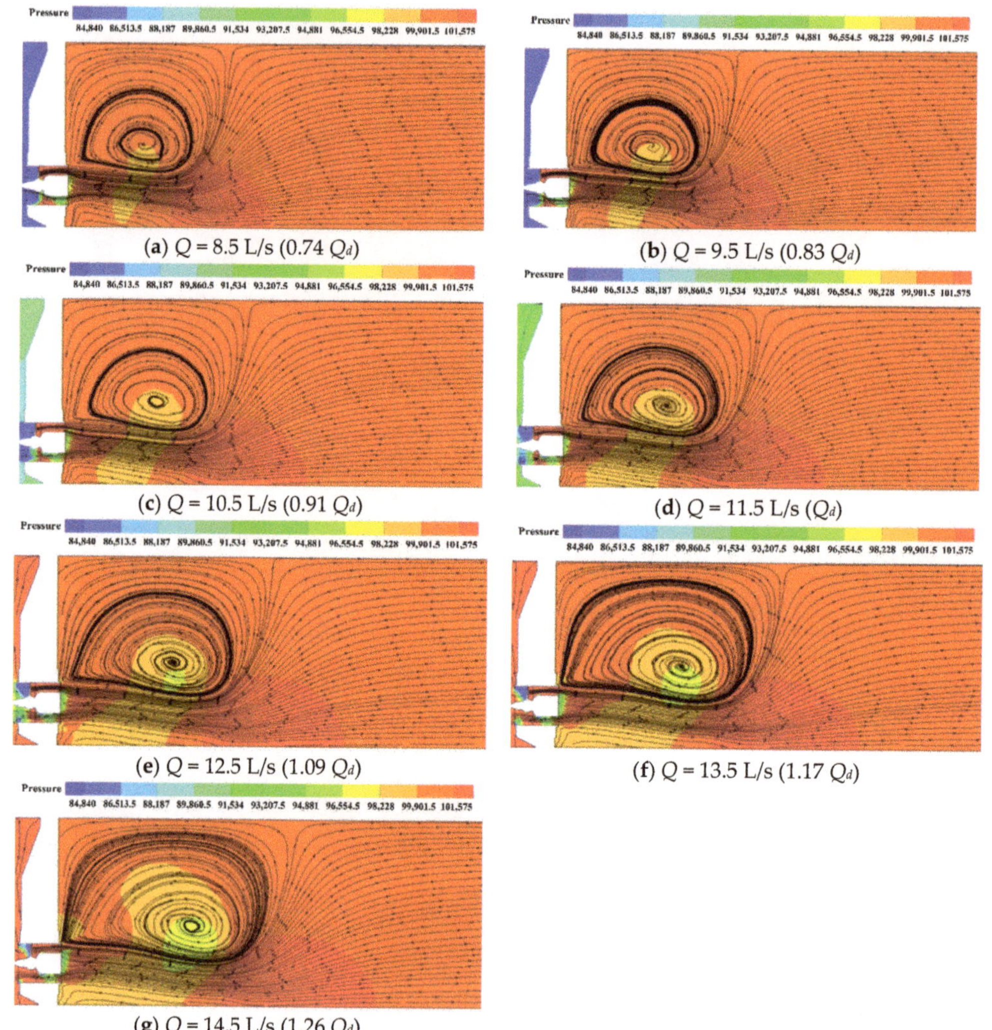

Figure 11. Streamline and pressure distribution at the pump gate outlet under different flow conditions.

As can be seen from Figure 11, at different flow rates, there is a backflow vortex above the pump gate outlet. With the increase of flow rate, the size of the vortex area also increases, and the center of the backflow vortex also shifted to the outlet of the outflow channel. Under the action of gravity, the fluid flows as a whole to the bottom of the pool to do offset flow, and due to the existence of vortex, the pressure at the outlet is low, and with the increase of vortex, the low pressure area also increases; pump gate outlet is 5~9 D at the bottom of the pool, and there is a semicircular piece of high pressure area, and due to the slope of the outflow channel, the pressure returns to normal level, and the streamline also began to level off.

3.3.2. Hydraulic Loss of Outlet Channel

According to the Formula (3) to calculate the hydraulic loss of the outflow channel, draw the hydraulic loss curve of the outflow channel under different flow conditions, as shown in Figure 12.

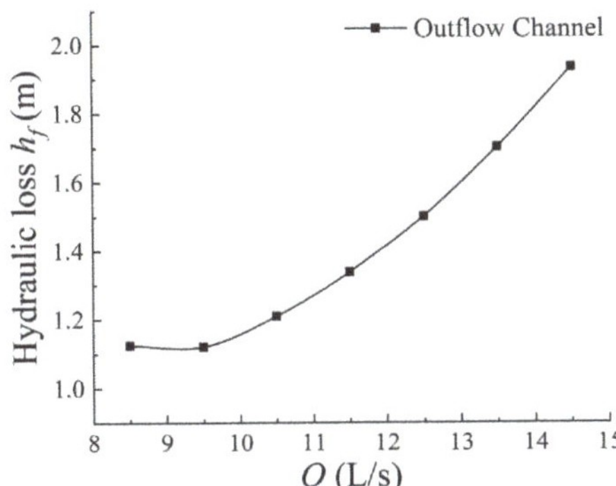

Figure 12. Hydraulic loss curve of outlet channel under different working conditions.

As can be seen from Figure 12, the hydraulic loss of the open outlet channel under different flow conditions is approximately parabolic with an open upward distribution, reaching a minimum value of 1.121 m at an inlet flow rate of 9.5 L/s (0.83 Q_d). With the increase of the inlet flow rate, the hydraulic loss also gradually increases, reaching a maximum value of 1.933 m at an inlet flow rate of 14.5 L/s (1.26 Q_d).

3.4. *Three-Dimensional Flow Regime Analysis*

3.4.1. Integrated Pump Gate Three-Dimensional Streamline and Characteristic Cross-Sectional Flow Rate

The characteristic section is shown schematically in Figure 13, the specific location is shown in Table 3, and the pump gate impeller diameter D = 60 mm.

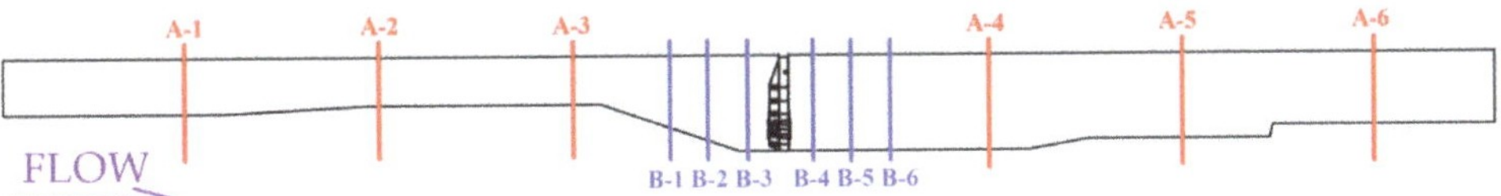

Figure 13. Schematic diagram of the characteristic cross section.

Setting characteristic sections A-1, A-2, A-3 in order at 10 D intervals from the impeller inlet. Setting characteristic sections A-4, A-5, A-6 in order at 10 D intervals from the guide vane outlet. Setting characteristic sections B-1, B-2, B-3 in order at distances 5 D, 3 D and D from the impeller inlet. Setting characteristic sections B-4, B-5, B-6 in order at distances D, 3 D and 5 D from the guide vane outlet.

The position of 5 D, 3 D, D from the impeller inlet and D, 3 D, 5 D from the guide vane outlet are selected to make cross sections, and the sections are shown in the characteristic sections B-1~B-6 in Figure 13, and the three-dimensional streamlines and axial flow velocity distribution of the integrated pump gate in the characteristic sections are shown in Figures 14–20.

Table 3. Summary of characteristic selections.

Characteristic Section Number	Distance from Impeller Inlet L_1	Distance from Guide Vane Outlet L_2
A-1	30 D	
A-2	20 D	
A-3	10 D	
A-4		10 D
A-5		20 D
A-6		30 D
B-1	5 D	
B-2	3 D	
B-3	D	
B-4		D
B-5		3 D
B-6		5 D

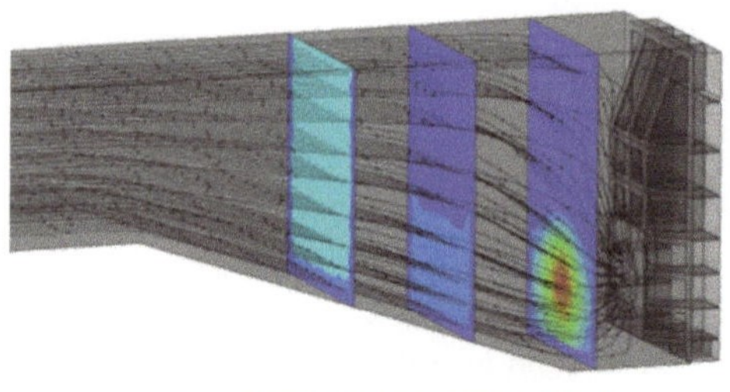
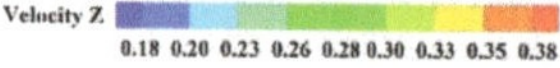

(**a**) Before the pump gate

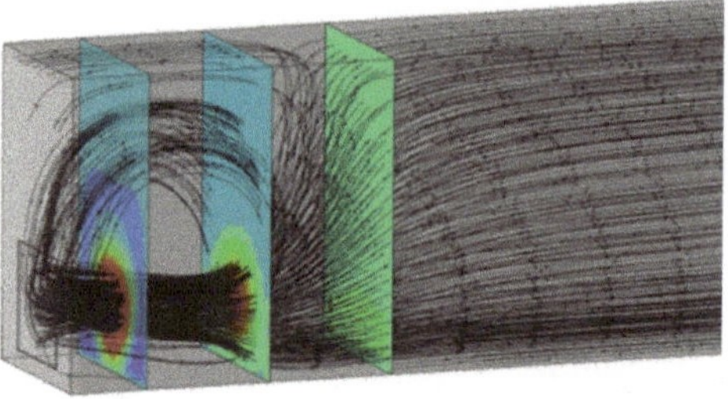

(**b**) After the pump gate

Figure 14. Q = 8.5 L/s (0.74 Q_d).

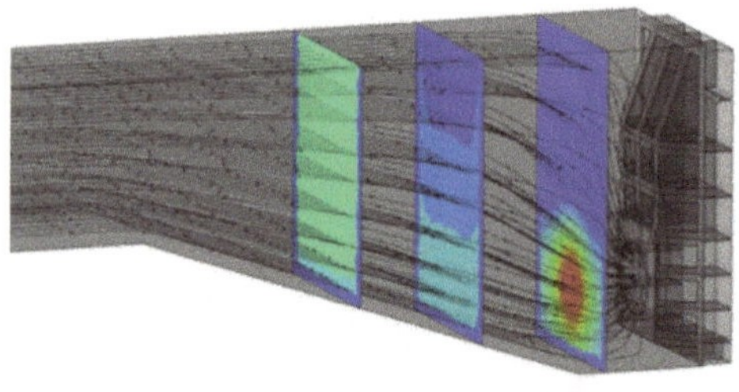

(**a**) Before the pump gate

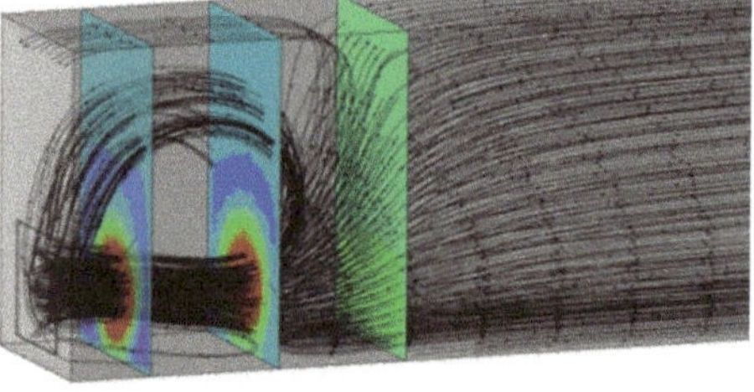

(**b**) After the pump gate

Figure 15. Q = 9.5 L/s (0.83 Q_d).

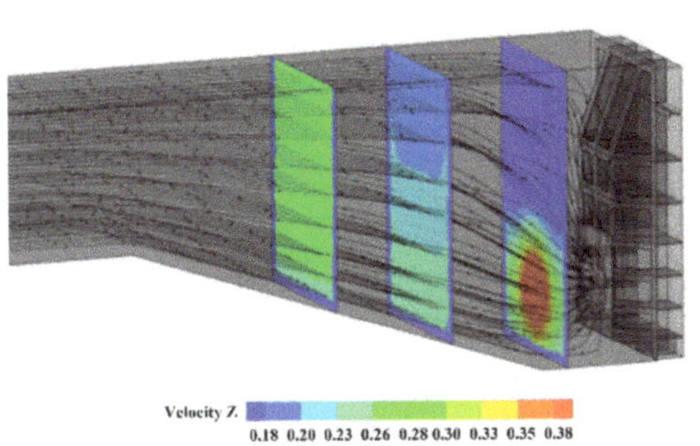

(**a**) Before the pump gate

Velocity Z
−0.50 −0.25 0.00 0.25 0.50 0.75 1.00 1.25 1.50

(**b**) After the pump gate

Figure 16. Q = 10.5 L/s (0.91 Q_d).

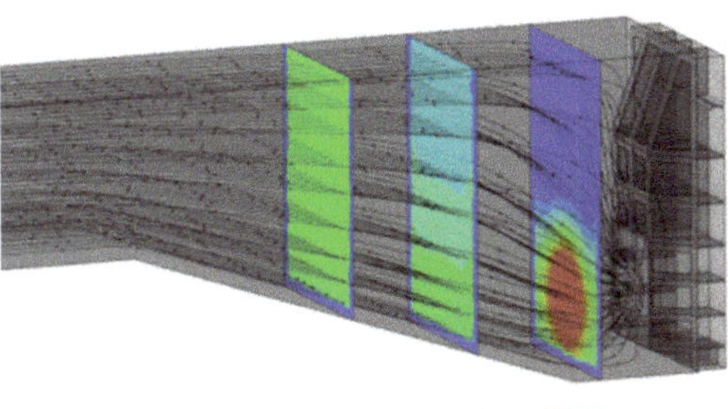

(**a**) Before the pump gate

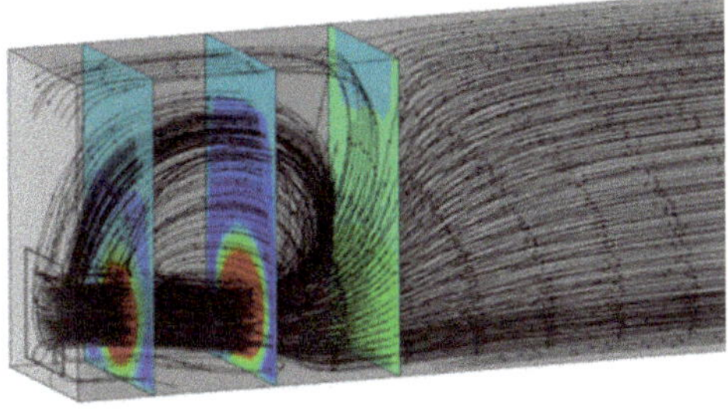

(**b**) After the pump gate

Figure 17. Q = 11.5 L/s (Q_d).

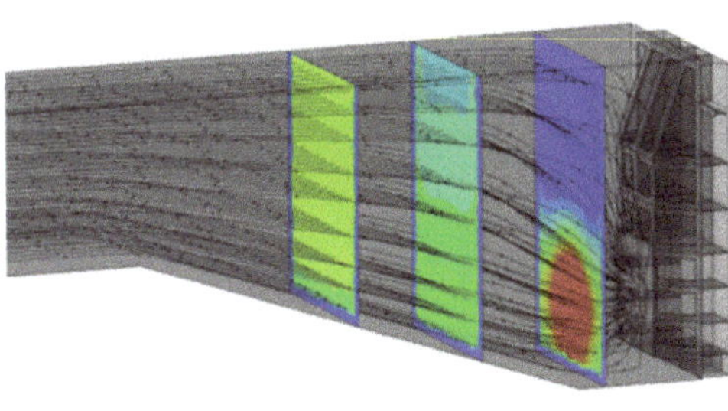

(**a**) Before the pump gate

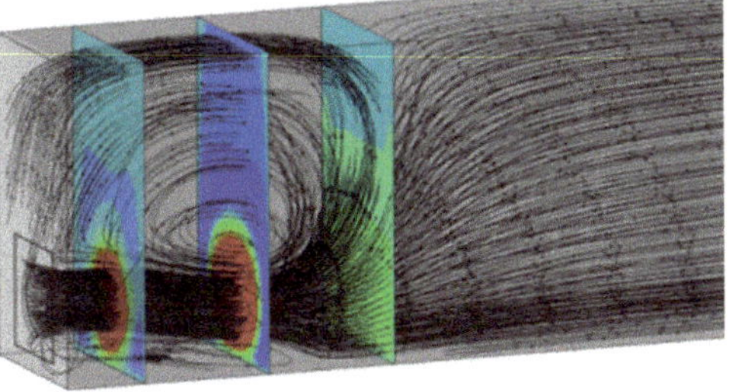

(**b**) After the pump gate

Figure 18. Q = 12.5 L/s (1.09 Q_d).

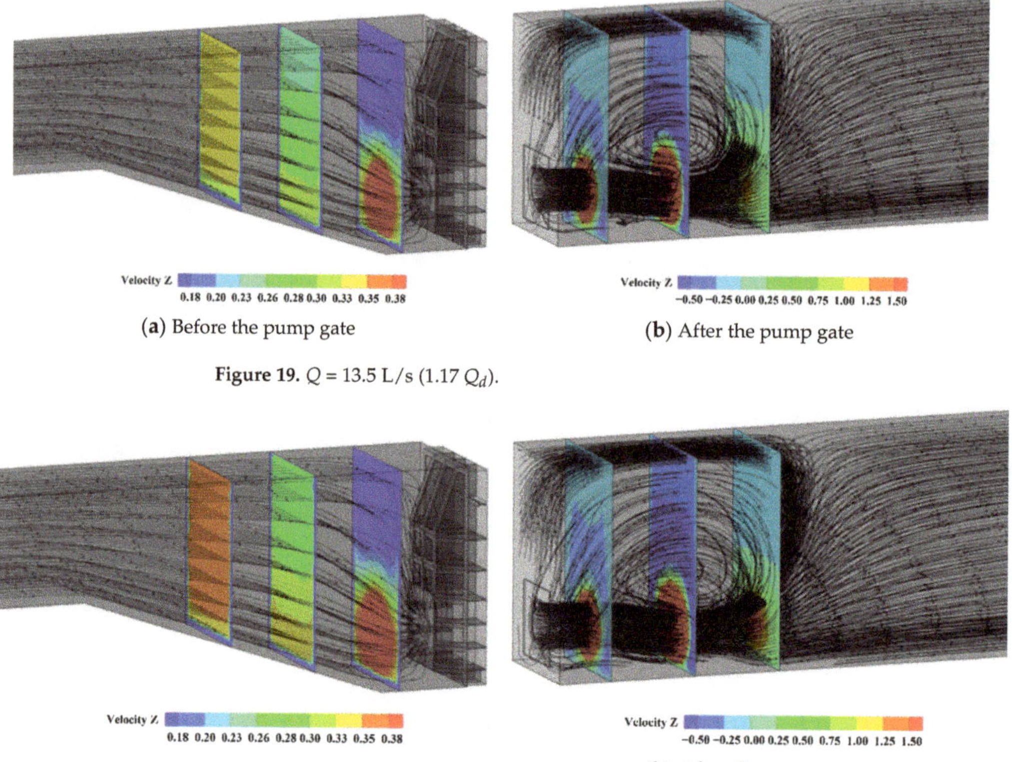

(**a**) Before the pump gate (**b**) After the pump gate

Figure 19. Q = 13.5 L/s (1.17 Q_d).

(**a**) Before the pump gate (**b**) After the pump gate

Figure 20. Q = 14.5 L/s (1.26 Q_d).

By comparing Figures 14 and 20, it can be seen that before the impeller inlet of the pump gate, with the increase of the inlet flow rate, there is no adverse flow pattern in the inlet channel, and the streamlines are uniformly contracted toward the impeller inlet. Comparing the axial velocity distribution clouds in the sections at different flow rates, it can be seen that the axial velocity in the same sections gradually increases with the increase of the inlet flow rate. At 5 *D* from the impeller inlet, the axial velocity distribution in the section is more uniform, and the velocity at the side wall is smaller due to the boundary layer effect. At 3 *D* from the impeller inlet, the velocity in the section gradually increases from top to bottom, which is mainly due to the slope that causes the water to collect at the bottom of the pool. At the impeller inlet *D*, the axial flow velocity is distributed outward in a circular pattern, and the high velocity area is located in the lower and middle part of the slice at the level of the impeller, while the flow velocity in the upper and middle part of the section is lower.

From the distribution of streamlines in the outlet section of Figures 14 and 20, it can be concluded that between 3 *D* and 5 *D* from the guide vane outlet, there is separation and stratification of the fluid caused by vortices, which is due to the opposite axial flow velocity of the streamlines. The region of high flow velocity within each section is located in the middle and lower part, and the axial flow velocity of the water in the middle and upper part is opposite to the flow velocity at the bottom of the flow channel due to the presence of vortices. At *D* and 3 *D* from the guide vane outlet, the axial flow velocity in the sections is distributed in a circular band, decreasing from the center line of the guide vane to the top and bottom of the sections. At 5 *D* from the guide vane outlet, when the flow rate is 8.5~13.5 L/s (0.74~1.17 Q_d), the axial flow velocity in the sections is approximately the same, and only the upper part of the sections is slightly lower. When the flow rate is 14.5 L/s (1.26 Q_d), there is still a circular region of high axial velocity in the section, which is caused by the fact that as the flow rate increases, the flow rate increases and the high velocity region also shifts toward the outlet of the outflow channel. At a flow rate of 8.5 L/s (0.74 Q_d), the area of the high flow velocity region in the slice at *D* from the guide vane outlet is larger than that in the slice at 3 *D* from the guide vane outlet, and the area of the high flow velocity region in the slice at 3 *D* from the guide vane outlet is larger than that in the slice at *D* from the guide vane outlet for the rest of the flow conditions, This is due to small flow conditions, small flow rate, short distance of high-speed water flow transmission at the outlet of pump gate, and fast dissipation of kinetic energy.

3.4.2. Uniformity of Axial Flow Velocity Distribution and Velocity-Weighted Average Angle of Each Characteristic Section under Design Conditions

Through the CFD-post post-processing interface, the velocity components of each unit in the characteristic section are derived, and the axial velocity components are extracted, and the axial velocity distribution uniformity and velocity-weighted average angle of each characteristic section are calculated in turn, according to Formulas (4) and (5), as shown in Table 4, where the characteristic sections B-4, B-5, and B-6 are closer to the guide vane outlet, and the water flow is not fully diffused, and their the axial velocity distribution uniformity and velocity weighted average angle are not very meaningful, so they are not selected. The axial velocity distribution uniformity and velocity-weighted average angle curves of each characteristic section under different flow conditions are plotted as shown in Figure 21.

Table 4. Uniformity of axial flow velocity distribution and velocity-weighted average angle for each characteristic section under design condition.

Evaluation Indicators	Section A-1	Section A-2	Section A-3	Section B-1	Section B-2	Section B-3	Section A-4	Section A-5	Section A-6
Uniformity of flow rate distribution *Vu* (%)	95.42	94.18	93.95	96.72	91.15	45.33	95.37	98.26	98.25
Velocity − weighted average angle $\overline{\theta}$(°)	89.93	89.71	89.54	79.47	75.30	62.01	87.95	89.67	89.96

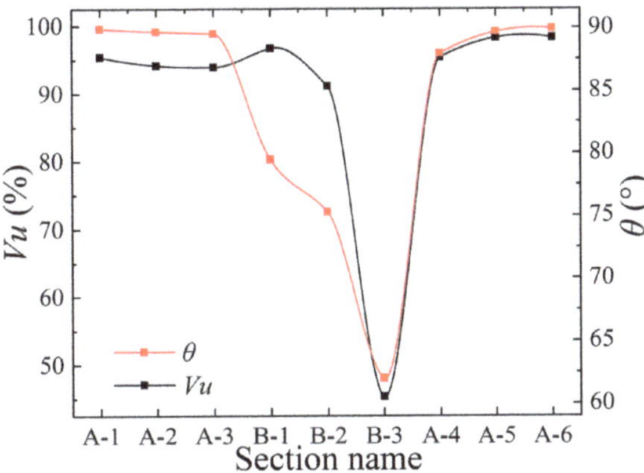

Figure 21. The uniformity of axial flow velocity distribution and velocity-weighted average angle curve of each characteristic section.

The more uniformity of axial flow velocity distribution is close to 100%, it means that the more uniform flow velocity in the section, by comparing the uniformity of axial flow velocity distribution in each characteristic section, it can be seen that the uniformity of axial flow velocity distribution in section B-2 and section B-3 is slightly lower under the design working condition. At section B-1, the inlet slope is more reasonable, and the uniformity of axial flow velocity distribution can reach 96.72%, as section B-2 and section B-3 are close to the impeller inlet. Influenced by the suction of the pump gate, the high-speed zone is concentrated in the axial direction of the impeller of the pump gate, the flow velocity gradually decreases in the axial direction in the shape of a ring around, the axial flow velocity distribution is first reduced to 91.15% (B-2), and then reduced to 45.33% (B-3), the closer to the pump gate inlet flow velocity uniformity is lower, from Figures 14–20 can also find this phenomenon. In the outlet channel, the uniformity of axial velocity distribution gradually increases, reaching 98.25% at section A-6, which indicates that the water flow diffusion is reasonable and the flow pattern affected by pump gate recovers quickly.

The closer the velocity-weighted average angle is to 90°, the better the isotropy of the water flow. By comparing the velocity-weighted average angle of each characteristic section, it can be seen that the velocity-weighted average angle decreases slightly at section A-1, section A-2, and section A-3, which is approximately a primary function relationship, and there is no obvious decrease. The velocity isotropy is better. At sections B-1, B-2, and B-3 near the impeller inlet, the velocity-weighted average angle decreases to 79.47°, 75.30°, and 62.01° respectively, and the closer to the pump gate, the smaller the velocity-weighted average angle is, which is also due to the influence of the high speed rotation of the pump gate, resulting in the velocity-weighted average angle in the section near the pump gate impeller is obviously different from other parts of the characteristic section. The velocity-weighted average angle in the section near the impeller of the pump gate is significantly different from that in other parts of the characteristic section. In the outflow channel, the velocity-weighted average angle gradually increases and reaches 89.96° at section A-6, indicating that the isotropy of the flow has been restored.

4. Internal Flow Characteristics Test Analysis

4.1. Introduction to the Pump Gate Test Rig

The total length of the test bench is about 7.5 m, the total width is about 1.8 m, and the diameter of the circulating pipe is 0.1 m. The test bench is made of transparent plexiglass, which can help visually and clearly observe the flow pattern of the inlet and outlet water channels of the pump gate, the position of the vortex, and the distribution of the bad flow pattern. The two-dimensional schematic, three-dimensional rendering and real objects of the test bench are shown in Figures 22 and 23 respectively, and the test measuring instruments and equipment are shown in Table 5.

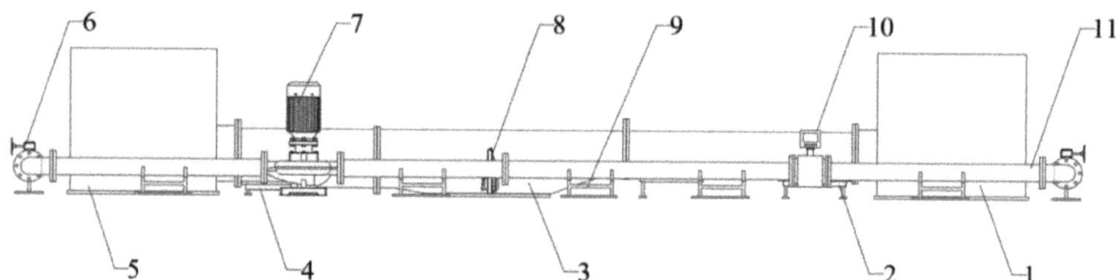

Figure 22. Schematic diagram of test bench: 1. water inlet tank; 2. water inlet channel support part; 3. open inlet and outlet channels; 4. outlet channel support part; 5. outlet water tank; 6. flange butterfly valve; 7. booster pump; 8. tested integrated pump gate; 9. pipe support; 10. electromagnetic flowmeter; 11. circulating pipeline.

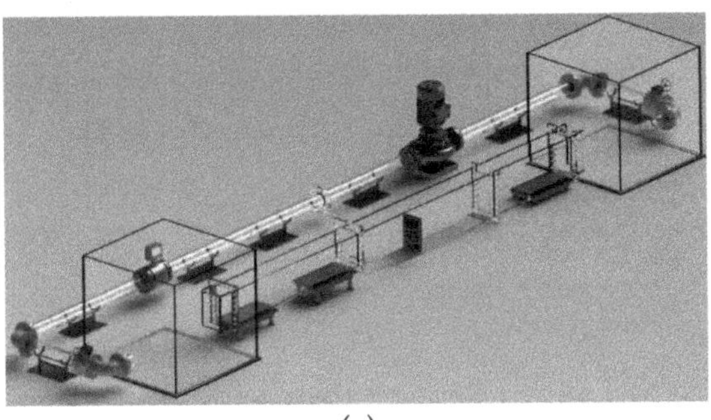

(a)

(b)

Figure 23. Three-dimensional rendering drawing and physical drawing of integral pump gate test stand: (**a**) Three-dimensional rendering drawing of integrated pump gate test stand; (**b**) physical drawing of integrated pump gate test stand.

Table 5. Main instruments and equipment for test data acquisition.

Measurement Items	Name of Measuring Equipment	Model	Scope of Work	Accuracy
Flow rate	Electromagnetic flow meters	ZEF-DN100	0~120 m^3/h	±0.5%
Rotational speed	Laser tachographs	DT-2234C	0.1~99,999 r/min	±0.05%
Flow pattern	High-speed cameras	OLYMPUS i-SPEED 3	2000 fpsFull resolutionMaximum 15,0000 fps	±1 μs

4.2. Test results and Analysis

4.2.1. Analysis of Flow Characteristics of Open Inlet and Outlet Water Channels

(1) Flow characteristics of open inlet channels

In this paper, the oscillation of the tracer red line is used to reflect the flow characteristics of the inlet channel. The flow patterns at positions A-1, A-2, and A-3 at the design flow rate $Q = 11.5$ L/s (Q_d) are selected for analysis by means of a high-speed camera, as shown in Figure 24.

From the tracer red lines in Figure 24, it can be seen that the water flow in the characteristic sections A-1, A-2, and A-3 is smooth, the streamline of the rear side wall of the characteristic section A-1 is evenly spaced, and the tracer red lines are parallel to each other, without cross winding. The tracer red line on the bottom rises slightly and floats upward, which is caused by the slope at the rear of section A-1, similar to the streamline state obtained in numerical simulation, the results of axial velocity distribution uniformity and velocity weighted average angle at this position can also show that the flow pattern here is smooth and the streamline is uniform. There is a horizontal flow passage between section A-2 and section A-3, and the streamline of this section remains horizontal, and the tracer red line does not appear disorderly and staggered, which is similar to the streamline diagram obtained by numerical simulation (Figure 6).

(2) Flow characteristics of open outlet channels

The flow characteristics of the water channel are reflected by the oscillating attitude of the tracer red line. The flow pattern is captured by a high-speed camera and selected for analysis at positions A-4, A-5, and A-6 at a design flow rate of $Q = 11.5$ L/s (Q_d), as shown in Figure 25.

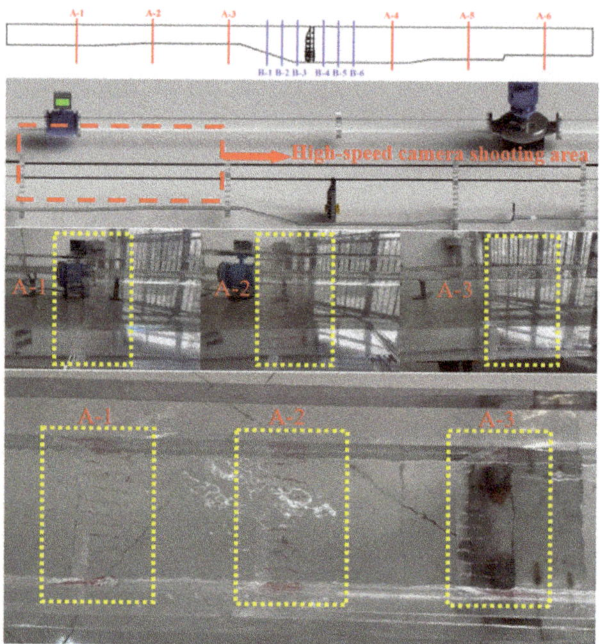

Figure 24. Flow pattern in an open inlet channel.

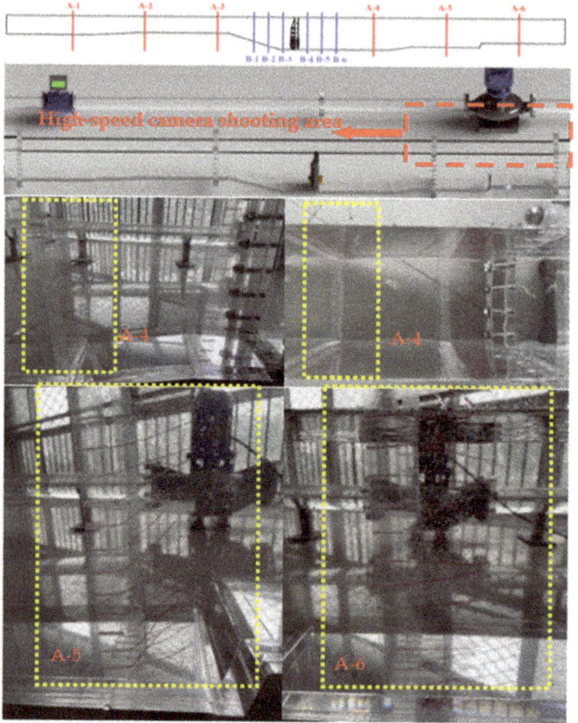

Figure 25. Flow pattern in an open outlet channel.

From the tracer red line in Figure 25, it can be seen that in the outlet channel, the streamline near the characteristic section A-4 is smooth, and the tracer red line on the wall and bottom slightly swings upward under the action of the rear slope of section A-4, which is similar to the phenomenon obtained by numerical simulation. At the characteristic section A-5, due to the existence of the outlet slope, the tracer red line swings up significantly, but before reaching the characteristic section A-6, the tracer red line has recovered to the level, which is similar to the streamline in the numerical simulation (Figure 10). This phenomenon can also be verified by comparing the weighted average angles of the two sections in Table 4 and Figure 21.

4.2.2. Analysis of the Flow Characteristics at the Inlet and Outlet of the Pump Gate Impeller

(1) Flow characteristics at the inlet of the pump gate impeller

As shown in Figure 26, by observing the tracer red line at the impeller inlet of the pump gate design flow Q = 11.5 L/s (Q_d), it can be found that the tracer red line converges to the impeller rotation center under the action of low pressure at the impeller inlet of the pump gate, gradually shrinks and swings evenly, there is no cross wrapping at the pump shaft, and the tracer red line at the upper part of the inlet channel faces the pump inlet, which is similar to the numerical calculation result (Figure 6), the slope of the inlet passage provides good inlet conditions for the impeller.

Figure 26. Flow pattern at pump gate impeller inlet.

(2) Flow characteristics at the outlet of the pump gate guide vane

As shown in Figure 27, the design flow rate of the pump gate Q = 11.5 L/s (Q_d) can be obtained through the test, and there is an obvious backflow vortex at the outlet of the guide vane, section B-4, the side wall against the lower and bottom tracer red line swing toward the outflow channel, the side wall against the upper tracer red line due to the existence of vortex, swing toward the inlet direction, section B-5, B-6 tracer streamline swing state similar to section B-4, the tracer red line oscillates upwards in the upper part of the outflow channel, while the tracer red line oscillates downwards in the outflow channel, and there is an obvious stratification of the water flow. This phenomenon can also be found through the previous numerical simulation results, which is due to the formation of backflow vortex caused by the opposite axial velocity of the water flow on the upper and lower side of the outlet channel of the pump gate.

Figure 27. Flow pattern at the outlet of the pump gate guide vane.

5. Conclusions

In this paper, the SST k-ω turbulence model is used to numerically simulate the integrated pump gate and verify its flow characteristics through experiments, the main conclusions are as follows.

(1) The numerical calculation results show that the efficiency of the integrated pump gate is 60.50% near the designed flow condition, the corresponding flow rate is 11.5 L/s, the head is 2.7569 m, the hydraulic loss of the intake channel is 0.064 m, and that of the outlet channel is 1.337 m.

(2) In this paper, through the axial velocity distribution and streamline of the inlet and outlet channels, combined with the pressure distribution at the inlet and outlet of the pump gate, the flow pattern of the integrated pump gate was analyzed, and it was concluded that the inlet flow pattern of the pump gate was uniform, and there was a large backflow vortex at the outlet of the pump gate. With the increase of the inlet flow, the vortex area increases and the center moves back.

(3) Integrated pump gate test results show that the streamline in the open inlet channel is smooth, parallel to each other, with no cross entanglement phenomenon. Impeller near the inlet, the tracer red line gradually converge to the impeller inlet, gradually contracted, uniform swing, no obvious bad flow pattern, vortex exists at the outlet of guide vane. In the middle and rear section of open outlet channel, the tracer red line gradually returns to parallel, and the streamline near the wall slightly swings upward at the outlet slope. The experimental phenomenon is similar to the numerical calculation result.

Author Contributions: Concept design, C.X. and W.X.; numerical calculation, A.F. and W.X.; experiment and data analysis, A.F. and F.S.; manuscript writing, C.X. and W.X. All authors have read and agreed to the published version of the manuscript.

Funding: This research was funded by Anhui Province Natural Science Funds for Youth Fund Project, grant number 2108085QE220. Key scientific research project of Universities in Anhui Province, grant number KJ2020A0103. Anhui Province Postdoctoral Researchers' Funding for Scientific Research Activities, grant number 2021B552. Anhui Agricultural University President's Fund, grant number 2019zd10. Stabilization and Introduction of Talents in Anhui Agricultural University Research Grant Program, grant number rc412008.

Institutional Review Board Statement: Not applicable.

Informed Consent Statement: Not applicable.

Data Availability Statement: Not applicable.

Conflicts of Interest: The authors declare no conflict of interest.

References

1. Zhao, Q. Analysis of the characteristics and application advantages of integrated pump gate. *Electromechanical Inf.* **2020**, 52–53. [CrossRef]
2. Zhou, M.; Meng, F.; Zhang, W. Development and application of an integrated pump gate. *Gen. Mach.* **2014**, 85–86.
3. Kentaro, F.; Yu, K.; Hiroaki, F. Validation of Inundation Damage Reduction by a Pump Gate with the New Type of Horizontal Axial Submersible Pump. *J. Disaster Res.* **2021**, *16*, 381–386.
4. Chen, W. *Numerical Simulation Study on Hydraulic Optimization of Integrated Pump Gate*; Yangzhou University: Yangzhou, China, 2021.
5. Shi, X.; Yan, G.; Dong, J.; Yang, Y. Safety study and structural optimization of vertical integrated pump gate. *Vib. Test. Diagn.* **2021**, *41*, 176–181+207.
6. Shi, X.; Yan, G.; Dong, J.; Sun, Y. Safety study and structural optimization of horizontal integrated pump gate. *Vib. Shock* **2021**, *40*, 42–47.
7. Wu, Y.; Zhao, H.; Liu, C. Based on The Gate Bottom Edge Structures Specific Numerical Simulation of Flow Pattern of Gate. *E3S Web Conf.* **2021**, *261*, 02018. [CrossRef]
8. Guo, M.; Chen, Z.; Lee, Y.; Choi, Y.D. Air Entrainment Flow Characteristics of Horizontal and Elbow Type Gate Pump-Sump Models. *KSFM J. Fluid Mach.* **2019**, *22*, 54–61. [CrossRef]
9. Chen, H.; Zheng, Y.; Zhou, D.; Shen, P.; Liu, H. Design and development of an eco-gate pump installation based on computational fluid dynamics. *Proc. Inst. Mech. Eng. Part C J. Mech. Eng. Sci.* **2017**, *231*, 2636–2649. [CrossRef]
10. Jiang, M. *ANSYS Workbench 19.0 Basic Introduction and Engineering Practice*; People's Post and Telecommunications Publishing Company: Beijing, China, 2019.
11. Shi, L.; Yuan, Y.; Jiao, H.; Tang, F.; Cheng, L.; Yang, F.; Jin, Y.; Zhu, J. Numerical investigation and experiment on pressure pulsation characteristics in a full tubular pump. *Renew. Energy* **2021**, *163*, 987–1000. [CrossRef]
12. Lu, Z.; Xiao, R.; Tao, R.; Li, P.; Liu, W. Influence of guide vane profile on the flow energy dissipation in a reversible pump-turbine at pump mode. *J. Energy Storage* **2022**, *49*, 104161. [CrossRef]
13. Launder, B.E.; Spalding, D.B. *Lectures in Mathematical Model of Turbulence*; Academic Press: London, UK, 1972.
14. Menter, F.R. *Zonal Two Equation k-ω Turbulence Models for Aerodynamic Flows*; AIAA-93-2906; AIAA: Preston, CA, USA, 1993.
15. Yang, F.; Zhang, Y.; Yao, Y.; Liu, C.; Li, Z.; Ahmed, N. Numerical and Experimental Analysis of Flow and Pulsation in Hump Section of Siphon Outlet Conduit of Axial Flow Pump Device. *Appl. Sci.* **2021**, *11*, 4941. [CrossRef]
16. Xie, C.; Tang, F.; Yang, F.; Zhang, W.; Zhou, J.; Liu, H. Numerical simulation optimization of axial flow pump device for elbow inlet channel. *IOP Conf. Ser. Earth Environ. Sci.* **2019**, *240*, 032006. [CrossRef]
17. Shi, L.; Zhang, W.; Jiao, H.; Tang, F.; Wang, L.; Sun, D.; Shi, W. Numerical simulation and experimental study on the comparison of the hydraulic characteristics of an axial-flow pump and a full tubular pump. *Renew. Energy* **2020**, *153*, 1455–1464. [CrossRef]
18. Zhang, W.; Tang, F.; Shi, L.; Hu, Q.; Zhou, Y. Effects of an Inlet Vortex on the Performance of an Axial-Flow Pump. *Energies* **2020**, *13*, 2854. [CrossRef]
19. Ahmed, N.; Fan, Y.; Yiqi, Z.; Tieli, W.; Mahmoud, H. Analysis of the Flow Pattern and Flow Rectification Measures of the Side-Intake Forebay in a Multi-Unit Pumping Station. *Water* **2021**, *13*, 2025. [CrossRef]
20. Sun, Z.; Yu, J.; Tang, F. The Influence of Bulb Position on Hydraulic Performance of Submersible Tubular Pump Device. *J. Mar. Sci. Eng.* **2021**, *9*, 831. [CrossRef]
21. Ji, D.; Lu, W.; Lu, L.; Xu, L.; Liu, J.; Shi, W.; Huang, G. Study on the Comparison of the Hydraulic Performance and Pressure Pulsation Characteristics of a Shaft Front-Positioned and a Shaft Rear-Positioned Tubular Pump Devices. *J. Mar. Sci. Eng.* **2021**, *10*, 8. [CrossRef]

Article

Study on Critical Velocity of Sand Transport in V-Inclined Pipe Based on Numerical Simulation

Rao Yao [1], Dunzhe Qi [2], Haiyan Zeng [3], Xingxing Huang [4], Bo Li [2], Yi Wang [2], Wenqiang Bai [2] and Zhengwei Wang [1,*]

1 Department of Energy and Power Engineering, Tsinghua University, Beijing 100084, China
2 Water Conservancy Project Construction Center of Ningxia Hui Autonomous Region, Yinchuan 750001, China
3 College of Water Resources and Civil Engineering, China Agricultural University, Beijing 100083, China
4 S.C.I.Energy, Future Energy Research Institute, Seidengasse 17, 8706 Zurich, Switzerland
* Correspondence: wzw@mail.tsinghua.edu.cn

Abstract: The Yellow River has a high sand content, and sand deposition in the pipelines behind the pumping station occurs from time to time. It is of great significance to reasonably predict the critical velocity of the small-angled V-inclined water transportation pipes. In this study, a Eulerian multiphase model was employed to simulate the solid–liquid two-phase flow. Based on the conservation of the sand transport rate, the critical velocity of the V-inclined pipe was predicted. The effects of simulated pipeline length, pipe inclination and particle size were investigated. The results show that when the simulated pipeline length reached a certain value, it did not affect the prediction of the critical velocity of the overall pipeline. The $\pm 2^{\circ}$ pipe inclination had a negligible effect on the critical velocity for transporting small-sized particles, but it led to the nonuniform and asymmetrical distribution of liquid velocity and sand deposition at the different cross-sections. As the particle size increased, the critical velocity also increased. However, the influence of particle size on the critical velocity is currently complicated, resulting in a large difference between numerical simulation and empirical formulas when transporting large-sized particles. Accurate prediction of critical velocity is important for long-distance water transportation pipelines to prevent sand deposition and reduce costs.

Keywords: V-inclined pipe; sand transport; critical velocity; numerical simulation

Citation: Yao, R.; Qi, D.; Zeng, H.; Huang, X.; Li, B.; Wang, Y.; Bai, W.; Wang, Z. Study on Critical Velocity of Sand Transport in V-Inclined Pipe Based on Numerical Simulation. *Water* **2022**, *14*, 2627. https://doi.org/10.3390/w14172627

Academic Editor: Bommanna Krishnappan

Received: 14 August 2022
Accepted: 23 August 2022
Published: 26 August 2022

Publisher's Note: MDPI stays neutral with regard to jurisdictional claims in published maps and institutional affiliations.

Copyright: © 2022 by the authors. Licensee MDPI, Basel, Switzerland. This article is an open access article distributed under the terms and conditions of the Creative Commons Attribution (CC BY) license (https://creativecommons.org/licenses/by/4.0/).

1. Introduction

The Yellow River in China is one of the rivers with the highest sand content in the world, containing a large proportion of small-sized suspended particles [1]. Due to the changes in the flow rate during the operation of the Yellow River long-distance water transportation pipelines, different degrees of sand deposition are produced in the pipelines. The characteristics of sand transport in the pipe can be described by the flow regime, which is divided into different velocities [2–7]. With a sufficiently high flow velocity, the sand can be completely suspended in the water, which can be considered as a homogeneous flow. However, if the flow velocity is low enough to reach a certain value, the sand separates from the water and the flow in the pipe becomes a heterogeneous flow. At an even lower flow velocity, the sand forms a moving bed at the bottom of the pipe and eventually a stationary deposition.

Scholars have conducted plenty of research on the velocities that delineate the different flow regimes, especially the critical velocity that triggers sand deposition at the bottom of the pipes [8–11]. Critical velocity is one of the most important parameters of pipelines, and it ensures the economic and safe operation of long-distance water transportation. In order to solve the sand deposition problem in the water transportation pipelines, the inflow velocity needs to be higher than the critical velocity. In practical engineering applications, the critical velocity law in the pipelines is extremely complex and affected by many factors.

There are three main factors that affect the critical velocity: (1) pipe characteristics: pipe diameter and wall roughness; (2) fluid characteristics: slurry velocity and sand content; and (3) particle characteristics: particle size and particle grading composition, etc.

The critical velocity corresponds to the transition of the particle motion state. Since scholars focus on different particle motion states, the definitions of critical velocity are also distinct. Wasp [12], Shook [13] and Kokpinar [14] considered the critical velocity as the lowest point on the head loss and velocity curve of the pipe. Azamathulla [15] believed that the critical velocity is the value at which the flow velocity changes from small to large until there is no sand deposition at all. Durand [16], Thomas [17], Graf [18] et al. defined the critical velocity as the velocity from large to small until some particles begin to deposit, i.e., the minimum velocity at which all particles can remain in motion. According to He [19], the critical velocity is the average velocity of the cross-section when obvious bedload movement occurs in the pipe. An [20] considers the critical velocity as the average velocity of the cross-section when sand is pushed forward linearly and slowly at the bottom of the pipe without pile deposition. Although the above definitions of critical velocity are different, they are judged by the occurrence of sand deposition at the bottom of the pipes.

At present, there are two methods for predicting the critical velocity: an empirical formula and numerical simulation. Disagreement on the definition of critical velocity, differences in experimental measurement methods and differences in the hydraulic parameters chosen for the calculations lead to different empirical formulas. The representative ones are the formulas of Durand, Wasp, Shook, Turian and He Wuquan. However, the structural form of these formulas and the parameters involved vary greatly and are not universal. Compared with the empirical formulas obtained from experiments, numerical simulation has the advantage of being less expensive and more adaptable. Therefore, it is necessary to study critical velocity using numerical simulation. With the development of CFD, the numerical simulation of solid–liquid two-phase flow has been developed, and three-dimensional numerical simulation has greater advantages in analyzing local pipe sections. Sajeev [21] verified the accuracy of solid–liquid two-phase flow by numerical simulation in comparison with experiments. Ling [22] used a simplified ASM model to simulate the low-concentration solid–liquid two-phase flow. Kaushal [23] performed numerical simulations of pipeline slurry flow with mono-dispersed fine particles at high concentrations using Mixture and Eulerian two-phase models and found that the Eulerian model gives more accurate predictions for both the pressure drop and concentration profiles. Januário [24] used the CFD-DEM method to analyze characteristics of slurry flow at different velocities and compared this with the experiments. Dabirian [25] numerically simulated the critical velocity and compared it with experiments to investigate the effects of parameters such as particle size and fluid viscosity on sand transport in horizontal pipelines and to illustrate the feasibility of numerical simulation predictions of critical velocity. Yang [26] investigated the transport and deposition characteristics of sand in the pipeline by means of the Eulerian multiphase flow model and examined the effects of inlet velocity, particle size and sand concentration.

Long-distance water transportation pipelines unavoidably follow undulating topography. Although the bending angle is small, it complicates the multiphase flow. Most studies have focused on the flow in horizontal pipes and other forms of inclined pipes. Therefore, it is of great significance to study the critical velocity of V-inclined pipe with a small bend angle. Al-lababidi [27] compared a horizontal pipe with a +5° upward-inclined pipe and found that the pipe inclination had a negligible effect on the critical velocity but a significant effect on the sand transport. Danielson [28] studied the sand transport in −1.35° and +4° upward-inclined pipes in liquid and liquid–gas flow and found that the pipe inclination had no significant effect on the sand transport in liquid. Dabirian [29] experimentally investigated the effect of parameters such as particle size, sand content and phase velocity on the three-phase flow of a +1.5° upward-inclined pipe. Conversely, Tebowei et al. [30] used the Eulerian–Eulerian two-fluid model to simulate the sand transport in the V-inclined pipeline. They found that the small-angled V-inclined pipe had a significant impact on sand disposition compared to the horizontal section, and the critical velocity was much higher at the downstream section of the V-inclined pipe. Nossair [31] experimentally studied a +3.6° upward-inclined pipe and found that a higher flow rate is required to eliminate sand deposition in small-angle upward-inclined pipe compared to horizontal pipe. Stevenson [32] reported that downward-inclined pipe is more prone to sand deposition than upward-inclined pipe. Wang [33] used a single variable to conduct experimental research on an inclined pipe. They found that there is an optimal angle between the horizontal and inclined pipes, which makes the critical velocity have a maximum value. They also modified Wasp's empirical formula to obtain the critical velocity of the inclined pipe. V-inclined pipes tend to be affected by the curvature of the pipe and require more attention compared to separate upward- and downward-inclined pipes.

In this paper, a three-dimensional numerical simulation of a small-angle V-slope water transportation pipe was performed to predict the critical velocity of the pipe based on the conservation of the sand transport rate. The accuracy of the numerical simulation was verified by comparing it with the empirical formula. The effect of simulated pipeline length, pipe inclination and particle size on the critical nonsiltation flow velocity was investigated by controlling a single variable.

2. Numerical Method

2.1. Governing Equations

In the solid–liquid two-phase flow, the water is the primary phase while the sand is the secondary phase. The Eulerian multiphase model was employed to predict the solid–liquid two-phase flow in the pipe [26,30,34,35].The model treats each phase as a continuous medium in time and space, existing in the same space and permeating each other. However, each phase has different volume fraction velocities, temperatures and densities. There is slip and interaction between phases. Momentum and continuity equations are solved for each phase. The following is the governing equation of the Eulerian model applicable to multiphase flow.

The description of multiphase flow as interpenetrating continua incorporates the concept of phasic volume fractions, denoted here by α_q .

$$\sum_{q=1}^{n} \alpha_q = 1 \tag{1}$$

The solid phase is represented by s, the liquid phase is represented by l, α_s and α_l is the volume fraction of the liquid phase and the solid phase respectively, then:

$$\alpha_s + \alpha_l = 1 \tag{2}$$

The conservation of the mass equation for phase q is:

$$\frac{\partial}{\partial t}(\alpha_q \rho_q) + \nabla \cdot (\alpha_q \rho_q \boldsymbol{v}_q) = \sum_{p=1}^{n} (\dot{m}_{pq} - \dot{m}_{qp}) \tag{3}$$

where ρ_q is the physical density of phase q, $\boldsymbol{v}_q$ is the velocity of phase q, $\dot{m}_{qp}$ characterizes the mass transfer from phase q to phase p, and $\dot{m}_{pq}$ characterizes the mass transfer from phase p to phase q, and t is time.

In this study, there was no mass transfer between the solid and liquid phases, so $\dot{m}_{qp}$ and $\dot{m}_{pq}$ are both 0, and Equation (3) was simplified as:

$$\frac{\partial}{\partial t}(\alpha_q \rho_q) + \nabla \cdot (\alpha_q \rho_q v_q) = 0 \tag{4}$$

The conservation of momentum equation for phase q is:

$$\frac{\partial}{\partial t}(\alpha_q\rho_q\boldsymbol{v}_q) + \nabla\cdot(\alpha_q\rho_q\boldsymbol{v}_q\boldsymbol{v}_q) = -\alpha_q\nabla p + \nabla\cdot\bar{\bar{\tau}}_q + \alpha_q\rho_q\boldsymbol{g} + \sum_{p=1}^{n}(\boldsymbol{R}_{pq} + \dot{m}_{pq}\boldsymbol{v}_{pq} - \dot{m}_{qp}\boldsymbol{v}_{qp}) + \left(\boldsymbol{F}_q + \boldsymbol{F}_{lift,q} + \boldsymbol{F}_{wl,q} + \boldsymbol{F}_{vm,q} + \boldsymbol{F}_{td,q}\right) \quad (5)$$

where $\bar{\bar{\tau}}_q$ is the qth-phase stress–strain tensor:

$$\bar{\bar{\tau}}_q = \alpha_q\mu_q\left(\nabla\boldsymbol{v}_q + \nabla\boldsymbol{v}_q{}^T\right) + \alpha_q\left(\lambda_q - \frac{2}{3}\mu_q\right)\nabla\cdot\boldsymbol{v}_q\bar{\bar{I}} \quad (6)$$

where $\boldsymbol{g}$ is the acceleration of gravity, μ_q and λ_q are the shear and bulk viscosity of phase q, $\bar{\bar{I}}$ is the unit tensor, $\boldsymbol{F}_q$ is an external body force, $\boldsymbol{F}_{lift,q}$ is a lift force, $\boldsymbol{F}_{wl,q}$ is a wall lubrication force, $\boldsymbol{F}_{vm,q}$ is a virtual mass force, $\boldsymbol{F}_{td,q}$ is a turbulent dispersion force, and p is the pressure shared by all phases. $\boldsymbol{R}_{pq}$ is an interaction force between phases. $\boldsymbol{v}_{pq}$ and $\boldsymbol{v}_{qp}$ are the interphase velocities.

This paper considered the effect of gravity on sand deposition. Only the interaction force was considered between phases, ignoring the forces that have less influence such as lift force and virtual mass force. Equation (5) can be simplified as:

$$\frac{\partial}{\partial t}(\alpha_q\rho_q v_q) + \nabla\cdot(\alpha_q\rho_q v_q v_q) = -\alpha_q\nabla p + \nabla\cdot\bar{\bar{\tau}}_q + \alpha_q\rho_q g + \boldsymbol{R}_{pq} \quad (7)$$

For solid–liquid two-phase flow, the interaction force uses the Wen-Yu model and satisfies $\boldsymbol{R}_{ls} = -\boldsymbol{R}_{sl}$:

$$\boldsymbol{R}_{ls} = \frac{3\alpha_s\alpha_l\rho_l|\boldsymbol{v}_s - \boldsymbol{v}_l|}{4d_s}\alpha_l{}^{-2.65}\cdot\frac{24}{\alpha_l Re_f}[1 + 0.15(\alpha_l Re_f)^{0.687}]\cdot(\boldsymbol{v}_l - \boldsymbol{v}_s) \quad (8)$$

where d_s is the diameter of the solid phase; and Re_f is the relative Reynolds number.

2.2. Turbulence Model

Three turbulence models are used to simulate turbulence in multiphase flows: the mixture turbulence model, the dispersed turbulence model and a per-phase turbulence model. These models were used by Kaushal [23], Li [34] and Ekambara [36], respectively. The choice of turbulence model is based on the importance of the second-phase turbulence. The mixture turbulence model uses mixture properties and mixture velocities, which are sufficient to capture important features of the turbulent flow. Therefore, it was also used in this study to reduce the computational overhead while meeting the accuracy requirements of the calculation.

The turbulence kinetic energy k and its rate of dissipation ε are obtained from the following transport equations:

$$\frac{\partial}{\partial t}(\rho_m k) + \nabla\cdot(\rho_m\boldsymbol{v}_m k) = \nabla\cdot\left(\left(\mu_m + \frac{\mu_{t,m}}{\sigma_k}\right)\nabla k\right) + G_{k,m} - \rho_m\varepsilon \quad (9)$$

$$[\frac{\partial}{\partial t}(\rho_m\varepsilon) + \nabla\cdot(\rho_m\boldsymbol{v}_m\varepsilon) = \nabla\cdot\left(\left(\mu_m + \frac{\mu_{t,m}}{\sigma_\varepsilon}\right)\nabla\varepsilon\right) + \frac{\varepsilon}{k}(C_{1\varepsilon}G_{k,m} - C_{2\varepsilon}\rho_m\varepsilon)] \quad (10)$$

where σ_k and σ_ε are the turbulent Prandtl numbers for k and ε, respectively. $C_{1\varepsilon}$ and $C_{2\varepsilon}$ are constants.

$G_{k,m}$, the generation of turbulence kinetic energy, is described as:

$$G_{k,m} = \mu_{t,m}\left(\nabla\boldsymbol{v}_m + (\nabla\boldsymbol{v}_m)^T\right) : \nabla\boldsymbol{v}_m \quad (11)$$

The turbulent viscosity $\mu_{t,m}$ for the mixture can be expressed as:

$$\mu_{t,m} = \rho_m C_\mu \frac{k^2}{\varepsilon} \tag{12}$$

where C_μ is a constant.

The mixture density ρ_m, velocity $\boldsymbol{v}_m$ and molecular viscosity μ_m are listed below:

$$\rho_m = \sum_{i=1}^{N} \alpha_i \rho_i \tag{13}$$

$$\boldsymbol{v}_m = \frac{\sum_{i=1}^{N} \alpha_i \rho_i \boldsymbol{v}_i}{\sum_{i=1}^{N} \alpha_i \rho_i} \tag{14}$$

$$\mu_m = \sum_{i=1}^{N} \alpha_i \mu_i \tag{15}$$

where α_i, ρ_i, μ_i and v_i are the volume fraction, density, viscosity and velocity of the ith phase, respectively.

2.3. Computational Domain and Grid-Independent Analysis

The typical small-angled V-inclined pipe in the pipeline irrigation project of the Yellow River irrigation area was used as the research object. It consisted of −2° downward-inclined pipe and +2° upward-inclined pipe, as shown in Figure 1. The prototype size was used for three-dimensional modeling. The pipe diameter and length were 2600 mm and 80 m, respectively. The sections denoted S1, S2, S3, S4 and S5 on the pipe, as shown in the figure, are the cross-sections where the predicted data were obtained for analysis. Sand deposition at different inflow velocities can be observed in these cross-sections. S3 was the section where the lowest point of the pipe was located. S1, S2, S4 and S5 were located at −20 m, −10 m, 10 m and 20 m, respectively, from the S3 section.

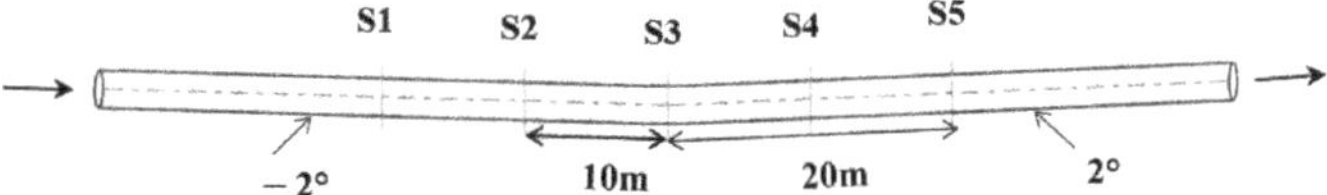

Figure 1. Schematic of the V-inclined pipe.

In the study of sand transport and deposition characteristics of multiphase flow, the grid is an important factor affecting the numerical simulation. The coarse mesh of 764,000 cells, the medium mesh of 1,487,600 cells, and the fine mesh of 1,855,600 cells were employed to analyze the sensitivity of the grid resolution to the numerical simulation results, as shown in Figure 2. Considering the influence of wall roughness on the sand deposition, five boundary layers were established along the surface with a growth factor of 1.2, and the height of the first layer from the wall was 5 mm.

The slurry velocity at the S3 cross-section was chosen to evaluate the effect of the grid number on the flow characteristics. Numerical simulations were carried out at an inlet volume fraction of 0.42% and an inlet velocity of 0.3 m/s. Table 1 shows the slurry velocity at the S3 cross-section for different grid numbers and the relative errors between them. Considering the computational time and accuracy, the final numerical simulation of the multiphase flow was carried out with a grid number of 1,487,600.

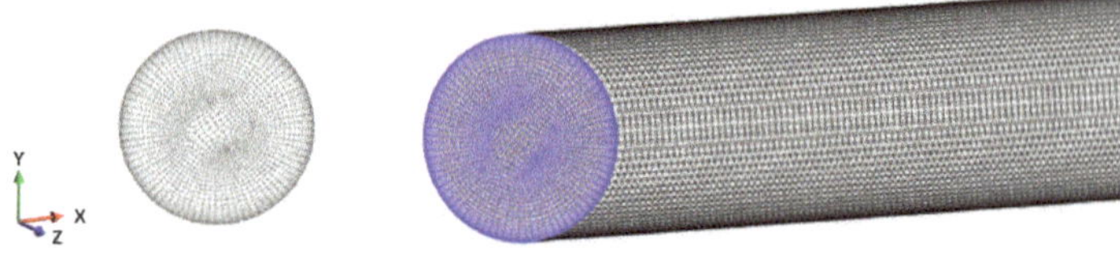

Figure 2. Grid structure of the pipe.

Table 1. Slurry velocity at the S3 cross-section under different grid numbers.

Scheme Number	Grid Number	Slurry Velocity (m/s)	Relative Error (%)
1	764,000	0.3029	0.66
2	1,487,600	0.3010	0.03
3	1,855,600	0.3009	0

2.4. Solution Strategies and Boundary Conditions

The ANSYS FLUENT software was used as a computational platform. The pressure and velocity equations were coupled using a phase-coupled SIMPLE algorithm. The continuity, momentum and turbulence equations were discretized by the second-order upwind scheme, while the volume fraction equation was discretized by the first-order upwind. Velocity and volume fraction of liquid and solid phases were assigned at the inlet condition of this pipe. The pressure was specified at the outlet condition of this pipe. The turbulence specification method used the intensity and hydraulic diameter. At the wall, the velocity was set to zero, which corresponds to the no-slip condition.

The literature shows that the particle size has a greater effect on critical velocity than bed roughness [37]. In the numerical simulation, the roughness of the pipe was not considered. The liquid phase was regarded as an incompressible fluid, and the physical properties of the solid phase were all constants, without considering the phase transformation. The shapes comprising the solid phase were treated as spherical particles of the same size. Table 2 shows parameters under different simulation conditions.

Table 2. Parameters under different simulation conditions.

Parameters	Horizontal Pipe	V-Inclined Pipe
		±2°
Grid cell size (m)	0.05	0.05
Pipe length L (m)	80	80/150/200
Pipe diameter D (mm)	2600	2600
Liquid density ρ ($kg \cdot m^{-3}$)	998.2	998.2
Sand density ρ_s ($kg \cdot m^{-3}$)	2300	2300
Particle size d (mm)	0.02	0.02\0.05\0.1
Sand content C_s ($kg \cdot m^{-3}$)	9.71	9.71
Sand volume concentration C_V (%)	0.42	0.42
Inflow velocities (m/s)	0.3–0.7	0.3–1.6

The numerical simulation method for solid–liquid two-phase flow used in this paper was validated by previous experiments [23,26]. The prediction of the critical velocity was achieved by setting different inflow velocities.

3. Results and Discussion

3.1. Comparison between Empirical Formula and Numerical Simulation

Many scholars have summarized the empirical formula of critical velocity in horizontal pipe through experiments. In this study, the critical velocity was predicted by numerical simulation based on the conservation of the sand transport rate. To verify the accuracy, a numerical simulation of the horizontal pipe was carried out and compared with the critical velocity obtained by the empirical formula.

3.1.1. Empirical Formula for Critical Velocity

There are differences in the definition and experimental measurement methods of critical velocity, resulting in different empirical formulas. The factors affecting the critical flow rate generally include sand content, particle size, pipe diameter, pipe roughness, etc. The representative empirical formulas are shown in Table 3.

Table 3. Empirical formula of critical velocity for horizontal pipe.

Reference	Empirical Formula
Durand [16]	$v_c = F_L\left(2gD\frac{\rho_s-\rho}{\rho}\right)^{\frac{1}{2}}$
Wasp [12]	$v_c = 3.28{C_V}^{0.243}\left(2gD\frac{\rho_s-\rho}{\rho}\right)^{\frac{1}{2}}\left(\frac{d}{D}\right)^{\frac{1}{6}}$
Shook [13]	$v_c = 2.43\frac{{C_V}^{\frac{1}{3}}}{{C_D}^{\frac{1}{4}}}\left(2gD\frac{\rho_s-\rho}{\rho}\right)^{\frac{1}{2}}$
He Wuquan [19]	$v_c = 1.8644K{C_W}^{0.2341}\left(gD\omega^2\frac{\rho_s-\rho}{\rho}\right)^{\frac{1}{4}}$

In Table 3 F_L is the modified Froude number when the solid particles appear to settle and thus deposit, which needs to be measured experimentally. C_D is the drag coefficient, ω is the settling velocity of particles, K is the correction factor, and the self-pressure pipe is taken as 1.05.

It can be seen from Table 3 that the empirical formulas based on experiments have similar structural characteristics, but the exponents and coefficients of each parameter are quite different. In most formulas, the critical velocity has an exponential relationship with the pipe diameter, and the critical velocity increases with the increase in the pipe diameter. At the same time, there is an exponential relationship between the critical velocity and sand content $v_c \propto {C_V}^m(0 < m < 1)$. Many studies have shown that the critical velocity increases with the increase in the sand content. When a certain limit is reached, the critical velocity decreases. There is a limit value between the critical velocity and the sand content. On the one hand, the increase in sand content inhibits the turbulent intensity of the liquid, and on the other hand, it increases the viscosity between particles and reduces the settling velocity of particles. However, the relationship between critical velocity and particle size is different. The particle size is not included in the formula of Shook. The critical velocity in the formula of Wasp is proportional to the particle size. Although the formula of He Wuquan does not contain particle size, the settling velocity in the formula is proportional to the particle size. It can be seen that the relationship between critical velocity and particle size is complicated.

3.1.2. Numerical Simulation for Critical Velocity

The empirical formula of critical velocity is influenced by experimental measurements as well as the measured parameters, and thus has poor applicability and reliability. Numerical simulation can be used to predict the critical velocity under different conditions, which is more economical.

The movement of sand in the pipelines of the Yellow River irrigation area is dominated by suspension movement, and sand deposition does not easily occur. However, due to the change in flow rate during operation, the sand deposition will occur when the flow rate is too low to maintain the suspension movement of sand. In this paper, critical velocity in the pipe was predicted based on the conservation of the sand transport rate. If the inlet and outlet mass flow rate of sand is basically equal, that is, the net mass flow rate is close to 0, the sand volume concentration no longer changes with time and sand in the pipe reaches the equilibrium between transport and deposition at that inflow velocity. In this paper, it was considered that on the premise of reaching the equilibrium between transport and deposition, the minimum inflow velocity at which the sand volume concentration does not increase with time is the critical velocity of the pipe. At the critical velocity, the sand moves slowly at the bottom of the pipe, but it will not accumulate in piles.

Three-dimensional numerical simulations under different inflow velocities for the horizontal pipe in Table 2 were carried out. Figures 3 and 4 show the change in net mass flow rate and sand volume concentration with time under different inflow velocities. It can be seen from Figure 3 that the net mass flow rate of sand was basically stable after the flow time of 1000 s. When the inflow velocity was greater than 0.3 m/s, and the net mass flow rate was stable near 0, that is, the sand transport at the inlet was equal to that at the outlet, and the equilibrium between transport and deposition was achieved. When the inflow velocity was 0.3 m/s, the net mass flow rate was stable around a certain value but still greater than 0, which indicates that the sand increases with time under this inflow velocity, and a part of sand is deposited at the bottom of the pipe. This conclusion can also be seen in Figure 4. When the inflow velocity was 0.3 m/s, the sand volume concentration increased gradually with time. However, when the inflow velocity increased to 0.4 m/s, the sand volume concentration did not increase with time. The stable sand volume concentration was slightly larger than the initial given concentration of 0.42%, indicating that there was still slight sand deposition. The sand at the bottom of the pipe slides or rolls on the bed surface, but does not affect transportation. In summary, the critical velocity of the horizontal pipe was predicted to be 0.4 m/s by numerical simulation.

The parameters of sand used in this paper come from real data from the pipeline irrigation project of the Yellow River irrigation area. The test sand sample of the He Wuquan formula is also from the Yellow River, so this empirical formula was selected to verify the accuracy of numerical simulation. Substituting the parameters of horizontal pipe in Table 2 into the He Wuquan formula, the critical velocity was found to be as 0.43 m/s. The results from the numerical simulation are in general agreement with this, which illustrates the accuracy of the numerical simulation in predicting the critical velocity.

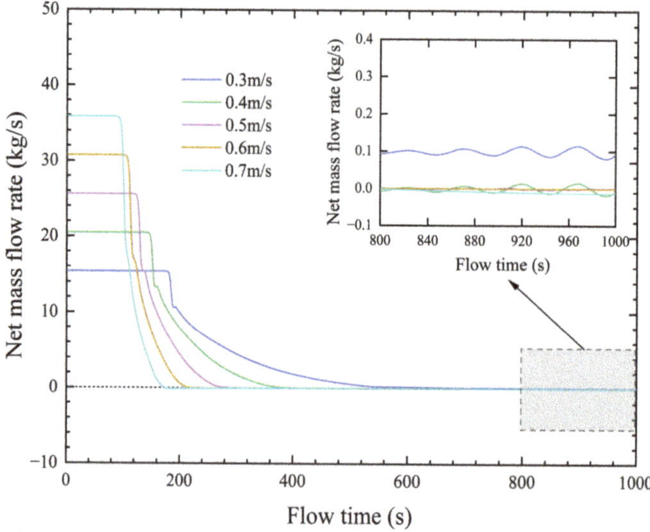

Figure 3. Net mass flow rate changes with flow time (horizontal pipe L = 80 m d = 0.02 mm).

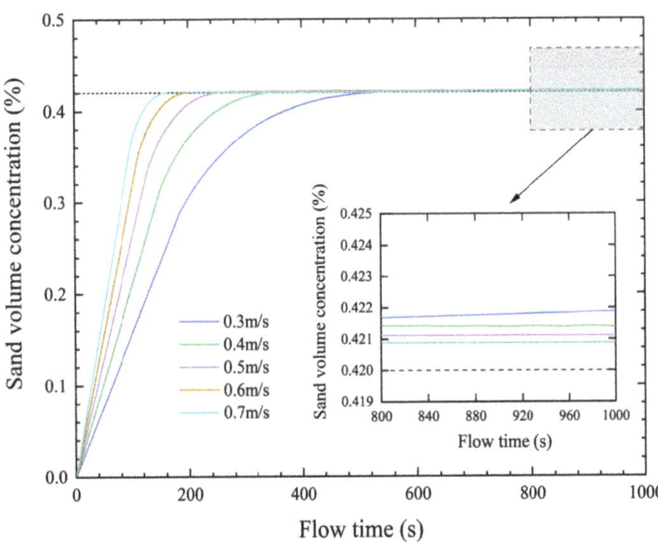

Figure 4. Sand volume concentration changes with flow time (horizontal pipe L = 80 m d = 0.02 mm).

3.2. Effect of Simulated Pipeline Length

The length of the pipeline in the irrigation project of the Yellow River irrigation area is as long as 40 km. Affected by computer performance, the three-dimensional numerical simulation of the entire pipeline cannot be carried out. Therefore, the effect of different simulated pipeline lengths on predicting the critical velocity of the pipe was investigated. The research object was the V-inclined pipe shown in Table 2. The simulated pipeline lengths were selected to be 80 m, 150 m and 200 m, respectively, and the particle size was set to 0.02 mm.

As shown in Figures 5–7, there was a change in the net mass flow rate with flow time under different simulated pipeline lengths of 80 m, 150 m and 200 m, respectively. It can be seen that the net mass flow rate of sand was basically stable after 1000 s, 1200 s and 1500 s for the simulated pipeline lengths of 80 m, 150 m and 200 m, respectively. When the inflow velocity was greater than 0.3 m/s, the net mass flow rate was stable near 0, reaching equilibrium between transport and deposition. When the inflow velocity was 0.3 m/s, the net mass flow rate was stable around a certain value, but still greater than 0. This shows that the sand in the pipe keeps increasing with time.

As shown in Figures 8–10, there was a change in sand volume concentration with flow time under different simulated pipeline lengths of 80 m, 150 m and 200 m, respectively. It can be seen that at the inflow velocity of 0.3 m/s, sand volume concentration had been increasing under the three simulated pipeline lengths, which is related to the fact that the net mass flow is not 0 after a certain period of flow, indicating that there is a continuous sand deposition in the V-inclined pipe. When the inflow velocity increased to 0.4 m/s, sand volume concentration no longer changed with time under the three simulated pipeline lengths. Therefore, the amount of sand deposition does not increase over time. The critical velocity of the V-inclined pipe at all three pipe lengths was 0.4 m/s.

In summary, the simulated pipeline length had almost no effect on the prediction of the critical velocity of the V-inclined pipe after a certain length. In addition, the stable sand volume concentration for different lengths at 0.4 m/s was slightly larger than the initially given value of 0.42%. Additionally, with the increase in simulated pipeline length, the sand volume concentration after stabilization was larger, which shows that simulated pipeline lengths have an effect on the amount of deposition and deposition forms.

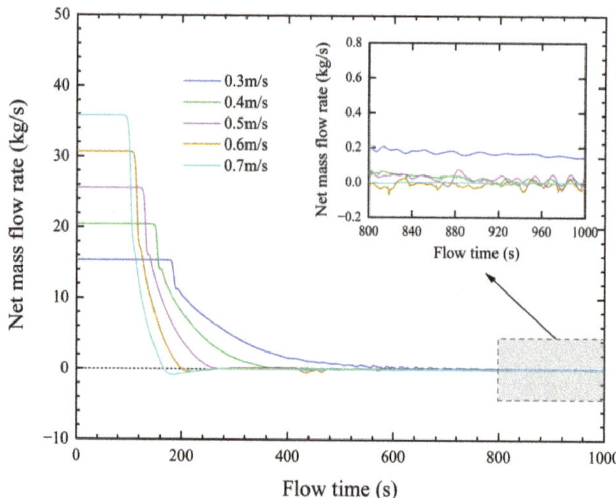

Figure 5. Net mass flow rate changes with flow time (V-inclined pipe L = 80 m d = 0.02 mm).

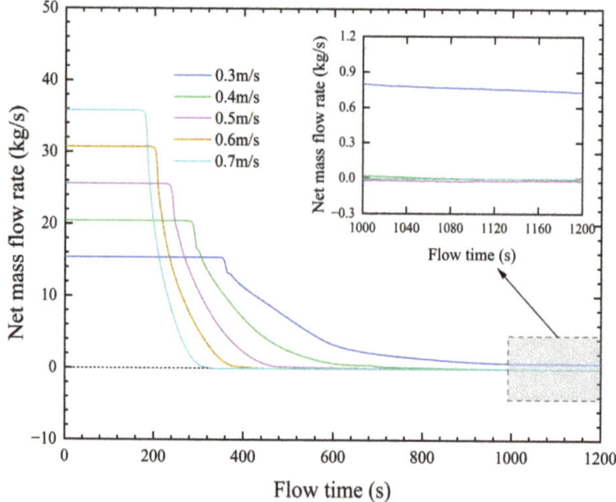

Figure 6. Net mass flow rate changes with flow time (V-inclined pipe L = 150 m d = 0.02 mm).

3.3. *Effect of Pipe Inclination*

The effect of pipe inclination on critical velocity can be obtained by comparing Figures 3 and 4 with Figures 5 and 8. It can be found that under the same conditions as other parameters, the critical velocity of $\pm 2^{\circ}$ V-inclined pipe and horizontal pipe when transporting particles with a size of 0.02 mm is 0.4 m/s for both.

Although the $\pm 2^{\circ}$ pipe inclination had no obvious effect on the critical velocity, the V-inclined pipe was different from the horizontal pipe in that the pipe curvature still had an impact on the flow and sand deposition. Figure 11 shows the liquid velocity at different cross-sections when the inflow velocity was 0.4 m/s. y is the height of the pipe. It can be seen that the liquid velocity of the section conformed to the distribution characteristics of high velocity in the center of the pipe and low velocity near the pipe wall. Pipe inclination had a certain influence on the liquid velocity of the cross-sections. The liquid velocity of the horizontal pipe was symmetrical similar to that of the central axis. However, the liquid velocity of the V-shaped inclined pipe presented a nonuniform and asymmetric distribution, especially in the upward pipe. The liquid velocity near the top of the pipe was higher than that near the bottom of the pipe.

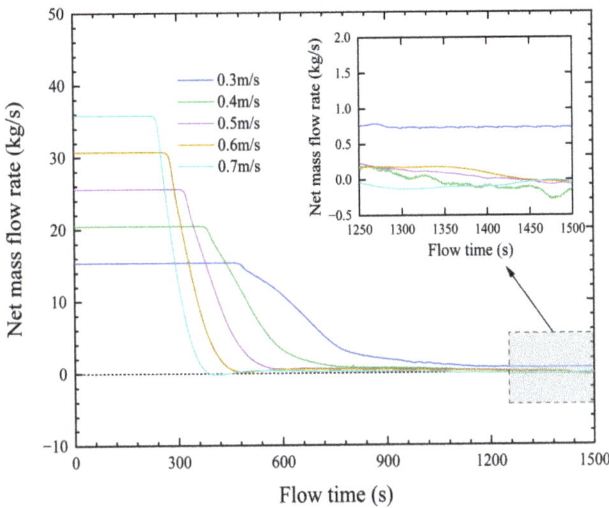

Figure 7. Net mass flow rate changes with flow time (V-inclined pipe L = 200 m d = 0.02 mm).

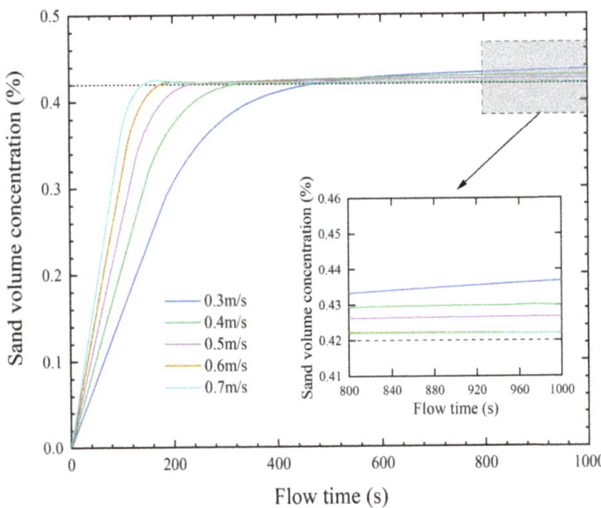

Figure 8. Sand volume concentration changes with flow time (V-inclined pipe L = 80 m d = 0.02 mm).

Figures 12 and 13 are the sand volume concentration contours at different cross-sections of the horizontal pipe and the V-inclined pipe, respectively. It can be seen that the distribution of sand in each cross-section of the horizontal pipe was basically the same. In the V-inclined pipe, there were obvious differences between the upward and downward pipes. Sand was mainly concentrated in the upward pipe. The sand deposition was the largest at the lowest cross-section of the pipe. The low-velocity zone produced by sand deposition had an impact on the liquid velocity distribution, and the low-velocity movement of sand squeezed the main flow, thus resulting in the uniform and asymmetric distribution in Figure 11b. The fundamental reason for these phenomena is that the force on the particles in the V-inclined pipe is different from that in the horizontal pipe. The effect of gravity on the upward and downward pipes is different. In the downward pipe section, the component force of gravity is in the same direction as the flow direction, which can promote the flow of sand, while in the upward pipe section, it acts as a resistance in the opposite direction.

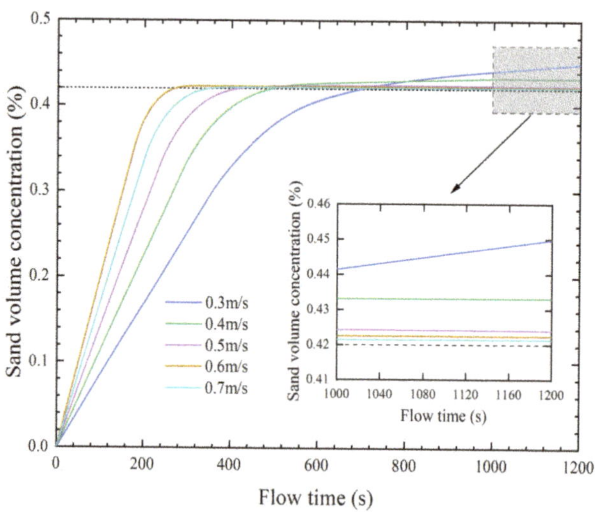

Figure 9. Sand volume concentration changes with flow time (V-inclined pipe L = 150 m d = 0.02 mm).

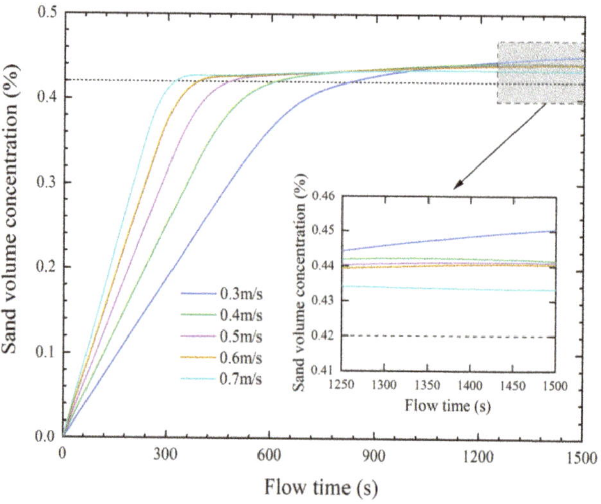

Figure 10. Sand volume concentration changes with flow time (V-inclined pipe L = 200 m d = 0.02 mm).

3.4. Effect of Particle Size

It can be seen from the empirical formula of the horizontal pipe that the relationship between the critical velocity and the particle size was complicated, and the critical velocity calculated by different formulas was very different. Therefore, the influence of different particle sizes on predicting the critical velocity of V-inclined pipe was investigated.

As shown in Figures 5, 14 and 15, there were changes in the net mass flow rate with the flow time under different particle sizes of 0.02 mm, 0.05 mm and 0.1 mm, respectively. The critical velocity was also predicted based on the conservation of the sand transport rate. It can be seen that the net mass flow rate was basically stable after the flow time of 1000 s under different particle sizes. When the flow velocity increased to 0.4 m/s, 1.1 m/s and 1.5 m/s respectively, the net mass flow rate was basically 0. At this time, sand volume concentration no longer changed with time. As shown in Figures 8, 16 and 17, the sand in the V-inclined pipe reached equilibrium between transport and deposition, and the amount of sand deposition did not increase with time. Therefore, the critical velocity of the pipeline under the particle sizes of 0.02 mm, 0.05 mm and 0.1 mm was 0.4 m/s, 1.1 m/s and 1.5 m/s, respectively.

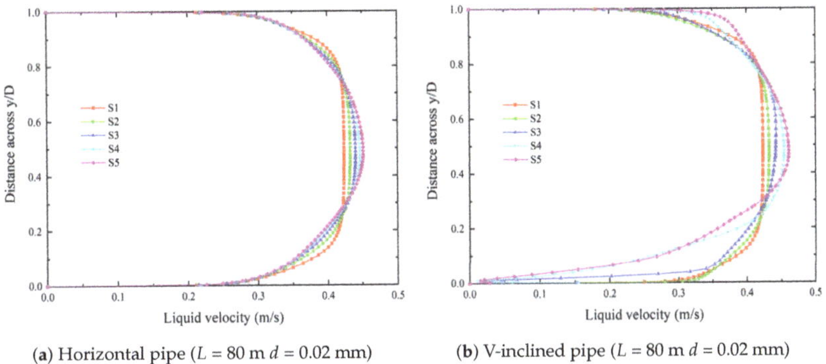

(**a**) Horizontal pipe (L = 80 m d = 0.02 mm) (**b**) V-inclined pipe (L = 80 m d = 0.02 mm)

Figure 11. Liquid velocity at different cross-sections (inflow velocity = 0.4 m/s).

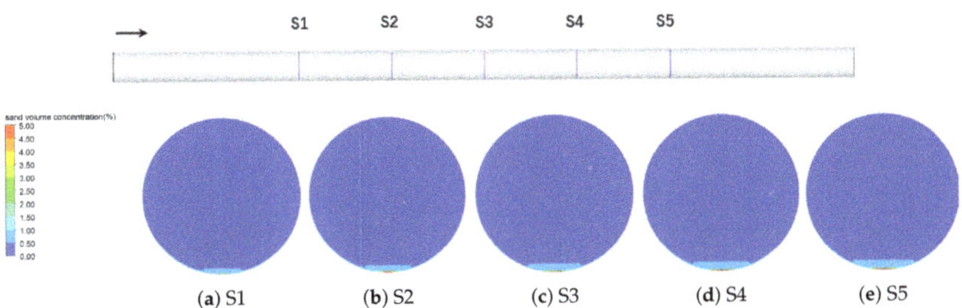

(**a**) S1 (**b**) S2 (**c**) S3 (**d**) S4 (**e**) S5

Figure 12. Sand volume concentration contours at different cross-sections of horizontal pipe (inflow velocity = 0.4 m/s).

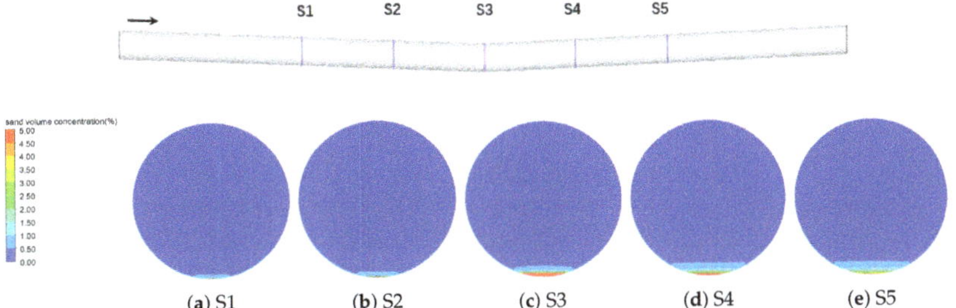

(**a**) S1 (**b**) S2 (**c**) S3 (**d**) S4 (**e**) S5

Figure 13. Sand volume concentration contours at different cross-sections of V-inclined pipe (inflow velocity = 0.4 m/s).

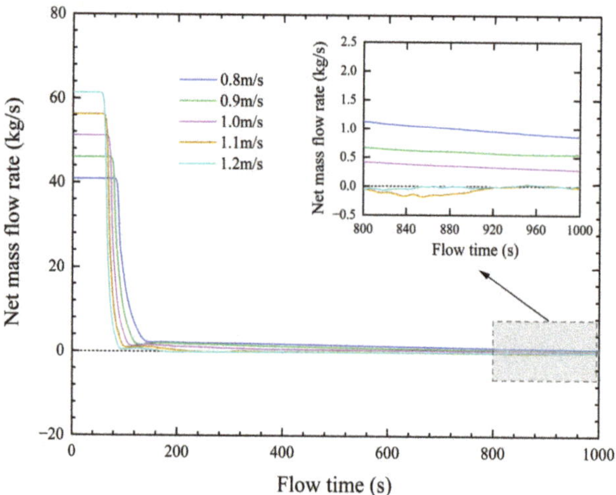

Figure 14. Net mass flow rate changes with flow time (V-inclined pipe L = 80 m d = 0.05 mm).

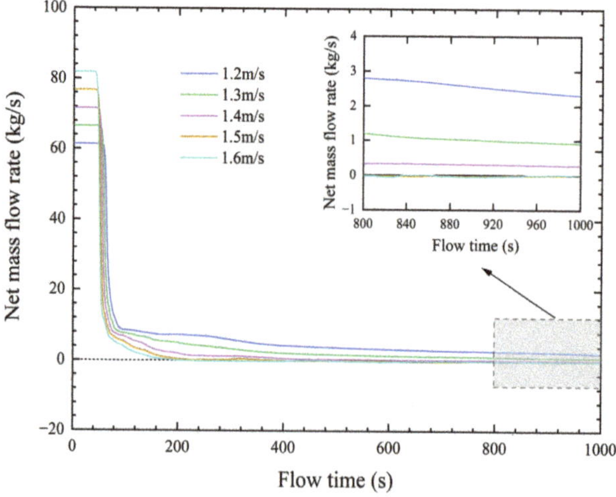

Figure 15. Net mass flow rate changes with flow time (V-inclined pipe L = 80 m d = 0.1 mm).

As the particle size increased, the critical velocity increased accordingly, which is consistent with the force of particles in the pipe. When the particles move in the pipe, they are affected by gravity and buoyancy in the vertical direction. The larger the particle size, the easier it is to deposit, and the greater the transporting velocity is required. There was no significant difference in the critical velocity in the V-inclined pipe when the horizontal pipe transport particle size was 0.02 mm, as mentioned in Section 3.3. The critical velocity under different particle sizes was calculated using the Wasp and He Wuquan formulas, which are related to the particle size given in Table 3. The comparison between empirical formulas and numerical simulation of the V-inclined pipe is shown in Table 4. When transporting smaller particles with particle sizes of 0.02 mm and 0.05 mm, the critical velocity calculated by the empirical formula and numerical simulation was basically the same. However, when transporting larger particles with a particle size of 0.1 mm, there was a significant difference. The reason for the difference is that, on the one hand, different empirical formulas have large differences in the prediction of critical velocity under the same particle size, and the inaccuracy of empirical formula prediction increases; on the other hand, the assumption of spherical particles is adopted in the numerical simulation, and only the drag force effect is considered between phases, which has a certain influence on the prediction of critical velocity.

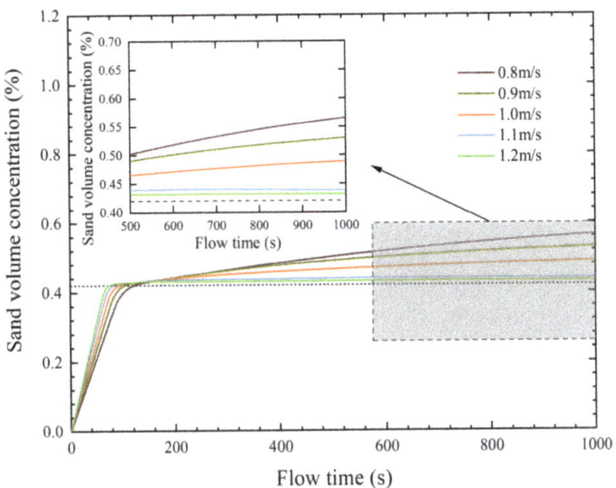

Figure 16. Sand volume concentration changes with flow time (V-inclined pipe L = 80 m d = 0.05 mm).

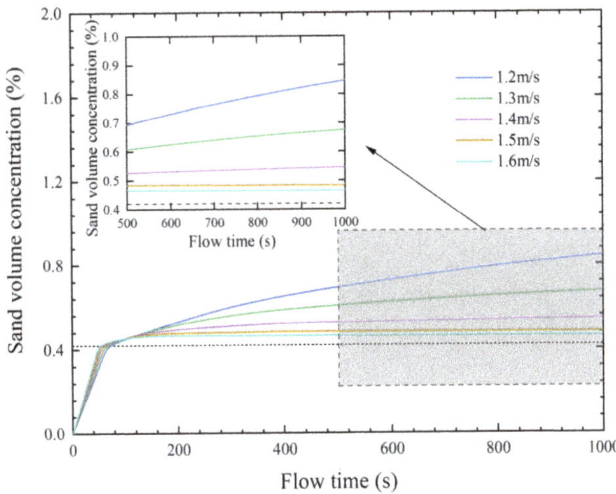

Figure 17. Sand volume concentration changes with flow time (V-inclined pipe L = 80 m d = 0.1 mm).

Table 4. Comparison between empirical formula and numerical simulation under different particle sizes.

Particle Size	Emprical Formula		Numerical Simulation
	Wasp	He Wuquan	
0.02 mm	0.99 m/s	0.43 m/s	0.4 m/s
0.05 mm	1.16 m/s	1.10 m/s	1.1 m/s
0.1 mm	1.30 m/s	2.05 m/s	1.5 m/s

4. Conclusions

Predicting critical velocity using numerical simulations can greatly reduce labor and cost. In this paper, the critical velocity of the $\pm 2^{\circ}$ V-inclined pipe was predicted using three-dimensional numerical simulation based on the conservation of the sand transport rate. The critical velocity predicted by the simulation of the horizontal pipe was basically consistent with the empirical formula, which verified the accuracy of the method. Numerical simulations of V-inclined pipe under different parameters were carried out, and the results show that:

(1) When the simulation length of the pipe reaches a certain value, it has no obvious effect on the prediction of the critical velocity of the V-inclined pipe. However, it will have an effect on the amount of deposition and deposition forms.

(2) Compared with the horizontal pipe, the $\pm 2^{\circ}$ pipe inclination has no obvious effect on the critical velocity of transporting 0.02 mm small-sized particles. In addition, the pipe inclination leads to the nonuniform and asymmetrical distribution of liquid velocity and sand deposition at different cross-sections. There are obvious differences between the upward and downward pipes. Sand is mainly concentrated in the upward pipe, and deposition is the largest at the lowest cross-section of the pipe. The effect of gravity on the particles in the downward and upward pipe is different.

(3) As the particle size increases, the critical velocity also increases. However, the effect of particle size on the critical velocity is complicated, resulting in a large difference between numerical simulation and empirical formula when transporting large-sized particles. On the one hand, the empirical formula for horizontal pipe may not be accurate in predicting the critical velocity of V-inclined pipe; on the other hand, the numerical simulation uses the assumption of spherical particles and only considers the drag effect between phases.

Author Contributions: Conceptualization, R.Y. and H.Z.; methodology, R.Y. and D.Q.; software, R.Y.; validation, X.H.; formal analysis, B.L.; investigation, Y.W.; resources, W.B.; writing—original draft preparation, R.Y.; writing—review and editing, Z.W. All authors have read and agreed to the published version of the manuscript.

Funding: This work was supported by the National Natural Science Foundation of China (No.: 51876099).

Institutional Review Board Statement: Not applicable.

Informed Consent Statement: Not applicable.

Data Availability Statement: Not applicable.

Acknowledgments: The authors gratefully acknowledge the financial support from the project Water conservancy science and technology in Ningxia Hui Autonomous Region [DSQZX-KY-01, DSQZX-KY-02].

Conflicts of Interest: The authors declare no conflict of interest.

Abbreviations

The following abbreviations are used in this manuscript:

CFD	computational fluid dynamics
ASM	algebraic slip mixture
DEM	discrete element method

References

1. He, W.Q.; Cai, M.K.; He, X.Y.; Zhang, C.D. Analysis and determination of critical non-silting velocity of muddy water conveyance pipelines in Yellow River irrigation districts. *Paiguan Jixie Gongcheng Xuebao/J. Drain. Irrig. Mach. Eng.* **2013**, *31*, 36–40.
2. Ardiclioglu, M.; Hadi, A.M.W.; Periku, E.; Kuriqi, A. Experimental and Numerical Investigation of Bridge Configuration Effect on Hydraulic Regime. *Int. J. Civ. Eng.* **2022**, *20*, 981–991. [CrossRef]
3. Daneshfaraz, R.; Aminvash, E.; Ghaderi, A.; Kuriqi, A.; Abraham, J. Three-Dimensional Investigation of Hydraulic Properties of Vertical Drop in the Presence of Step and Grid Dissipators. *Symmetry* **2021**, *13*, 895. [CrossRef]

4. Daneshfaraz, R.; Norouzi, R.; Abbaszadeh, H.; Kuriqi, A.; Di Francesco, S. Influence of Sill on the Hydraulic Regime in Sluice Gates: An Experimental and Numerical Analysis. *Fluids* **2022**, *7*, 244. [CrossRef]
5. Dasineh, M.; Ghaderi, A.; Bagherzadeh, M.; Ahmadi, M.; Kuriqi, A. Prediction of Hydraulic Jumps on a Triangular Bed Roughness Using Numerical Modeling and Soft Computing Methods. *Mathematics* **2021**, *9*, 3135. [CrossRef]
6. Lu, Z.; Xiao, R.; Tao, R.; Li, P.; Liu, W. Influence of guide vane profile on the flow energy dissipation in a reversible pump-turbine at pump mode. *J. Energy Storage* **2022**, *49*, 104161. [CrossRef]
7. Lu, Z.; Zhang, F.; Jin, F.; Xiao, R.; Tao, R. Influence of the hydrofoil trailing-edge shape on the temporal-spatial features of vortex shedding. *Ocean Eng.* **2022**, *246*, 110645. [CrossRef]
8. Oudeman, P. Sand Transport and Deposition in Horizontal Multiphase Trunklines of Subsea Satellite Developments. In *Offshore Technology Conference*; OnePetro: New York, NY, USA, 1992.
9. Bello, K.; B. Oyeneyin; G. Oluyemi. Minimum Transport Velocity Models for Suspended Particles in Multiphase Flow Revisited. In Proceedings of the SPE Annual Technical Conference and Exhibition, Denver, CO, USA, 30 October–2 November 2011; Volume 4.
10. Doron, P.; Barnea, D. Pressure Drop and Limit Deposit Velocity for solid–liquid Flow in Pipes. *Chem. Eng. Sci.* **1995**, *50*, 1595–1604. [CrossRef]
11. Doan, Q.; Ali, S.M.; George, A.E.; Oguztoreli, M. Simulation of sand transport in a horizontal well. In Proceedings of the International Conference on Horizontal Well Technology, Calgary, AB, Canada, 18–20 November **1996**; pp. 581–593.
12. Wasp, E.; Kenny, J.; Gandhi, R. Solid liquid flow—Slurry pipeline transportation. *Trans. Tech. Publ. Rockport MA* **1977**, *43*, 101–109.
13. Shook, C. Pipelining Solids: The design of short distance pipelines. *Proceedings from the Symposium on Pipeline Transport of Solids*; Canadian Society for Chemical Engineering: Toronto, ON, Canada, 1969.
14. Kokpinar, M.; Gogus, M. Critical Flow Velocity in Slurry Transporting Horizontal Pipelines. *J. Hydraul. Eng.* **2001**, *127*, 763–771. [CrossRef]
15. Azamathulla, H.M.; Ahmad, Z. Estimation of Critical Velocity for Slurry Transport through Pipeline Using Adaptive Neuro-Fuzzy Interference System and Gene-Expression Programming. *J. Pipeline Syst. Eng. Pract.* **2013**, *4*, 131–137. [CrossRef]
16. Durand, R. Basic relationships of the transportation of solids in pipes-experimental research. In Proceedings of the International Association of Hydraulic Research, Minneapolis, MN, USA, 1–4 September, 1953; pp. 89–103.
17. Thomas, D. Transport Characteristics of Suspensions: Application of Different Rheological Models to Flocculated Suspension Data. In *Progress in International Research on Thermodynamic and Transport Properties*; Academic Press: Cambridge, MA, USA, 1962; pp. 704–717.
18. Graf, W.; Robinson, M.; Yucel, O. Critical velocity for solid–liquid mixtures. In Proceedings of the 1st Conference on Hyd Transport of Solids in Pipes, BHRA Fluid Engineering, Cranfield, UK, 1970.
19. He, W.; Wang, Y.; Zhang, Y. Experimental on Muddy Water Delivery for Irrigation in Low-pressure Pipeline System. *J. Shenyang Agric. Univ.* **2007**, *38*, 98–101.
20. Jie, A.N.; Zong, Q.; Tang, H. Experimental Study on Non-depositing Critical Velocity of Muddy Water Delivery in Low-pressure Pipeline System. *J. Shihezi Univ. (Natl. Sci.)* **2012** , *30*, 83–86.
21. Sajeev, S.; Mclaury, B.; Shirazi, S. Critical Velocities for Particle Transport from Experiments and CFD Simulations. *Int. J. Environ. Ecol. Eng.* **2017**, *11*, 538–542.
22. Ling, J.; Skudarnov, P.V.; Lin, C.X.; Ebadian, M.A. Numerical investigation of liquid-solid slurry flows in a fully developed turbulent flow region. *Int. J. Heat Fluid Flow* **2003**, *24*, 389–398. [CrossRef]
23. Kaushal, D.R.; Thinglas, T.; Tomita, Y.; Kuchii, S.; Tsukamoto, H. CFD modeling for pipeline flow of fine particles at high concentration. *Int. J. Multiph. Flow* **2012**, *43*, 85–100. [CrossRef]
24. Januário, J.; Maia, C. CFD-DEM Simulation to Predict the Critical Velocity of Slurry Flows. *J. Appl. Fluid Mech.* **2020**, *13*, 161–168. [CrossRef]
25. Dabirian, R.; Arabnejad Khanouki, H.; Mohan, R.S.; Shoham, O. Numerical Simulation and Modeling of Critical Sand-Deposition Velocity for Solid/Liquid Flow. *SPE Prod. Oper.* **2018**, *33*, 866–878. [CrossRef]
26. Yang, Y.; Peng, H.; Wen, C. Sand Transport and Deposition Behaviour in Subsea Pipelines for Flow Assurance. *Energies* **2019**, *12*, 4070. [CrossRef]
27. Al-lababidi, S.; Yan, W.; Yeung, H. Sand Transportations and Deposition Characteristics in Multiphase Flows in Pipelines. *J. Energy Resour. Technol.* **2012**, *134*, 034501. [CrossRef]
28. Danielson, T. Sand Transport Modeling in Multiphase Pipelines. In Proceedings of the Offshore Technology Conference, Houston, TX, USA, 30 April–3 May 2007.
29. Dabirian, R.; Mohan, R.; Shoham, O.; Kouba, G. Sand Transport in Slightly upward-inclined Multiphase Flow. *J. Energy Resour. Technol.* **2018**, *140*, 072901. [CrossRef]
30. Tebowei, R.; Hossain, M.; Islam, S.Z.; Droubi, M.G.; Oluyemi, G. Investigation of sand transport in an undulated pipe using computational fluid dynamics. *J. Pet. Sci. Eng.* **2018**, *162*, 747–762. [CrossRef]
31. Nossair, A.; Rodgers, P.; Goharzadeh, A. Influence of Pipeline Inclination on Hydraulic Conveying of Sand Particles. In Proceedings of the ASME International Mechanical Engineering Mechanical Engineering Congress and Exposition, Houston, TX, USA, 9–15 November 2012, Volume 7.
32. Stevenson, P.; Thorpe, R.B. Towards understanding sand transport in subsea flowlines. In Proceedings of the 9th International Conference on Multiphase '99: Cannes, Paris, France, 16–18 June 1999; pp. 583–594.

33. Wang, J.; Li, Y.; Pan, L.; Lai, Z.; Jian, S. Study of the Sediment Transport Law in a Reverse-Slope Section of a Pressurized Pipeline. *Water* **2020**, *12*, 3042. [CrossRef]
34. Li, M.Z.; He, Y.P.; Liu, Y.D.; Huang, C. Hydrodynamic simulation of multi-sized high concentration slurry transport in pipelines. *Ocean Eng.* **2018**, *163*, 691–705. [CrossRef]
35. Parkash, O.; Kumar, A.; Sikarwar, B.S. Analytical and comparative investigation of particulate size effect on slurry flow characteristics using computational fluid dynamics. *J. Therm. Eng.* **2021**, *7*, 220–239. [CrossRef]
36. Ekambara, K.; Sanders, R.S.; Nandakumar, K.; Masliyah, J.H. Hydrodynamic Simulation of Horizontal Slurry Pipeline Flow Using ANSYS-CFX. *Ind. Eng. Chem. Res.* **2009**, *48*, 8159–8171. [CrossRef]
37. Alihosseini, M.; Thamsen, P.U. Analysis of sediment transport in sewer pipes using a coupled CFD-DEM model and experimental work. *Urban Water J.* **2019**, *16*, 259–268. [CrossRef]

Article

Energy Characteristics and Internal Flow Field Analysis of Centrifugal Prefabricated Pumping Station with Two Pumps in Operation

Chuanliu Xie [1,*], Zhenyang Yuan [1], Andong Feng [1], Zhaojun Wang [2] and Liming Wu [2]

1 College of Engineering, Anhui Agricultural University, Hefei 230036, China
2 Suqian Branch of Jiangsu Water Source Company of South to North Water Diversion, Suqian 223800, China
* Correspondence: xcltg@ahau.edu.cn

Abstract: In order to study the hydraulic performance and internal flow field of dual pumps in centrifugal prefabricated pumping station under operation conditions, this paper carried out a numerical calculation based on CFD software for dual pumps in a centrifugal prefabricated pumping station under different flow conditions and verified the internal flow field through test. The results show that the efficiency of centrifugal prefabricated pumping station under design conditions (Q_d = 33.93 m^3/h) is 63.96%, the head is 8.66 m, the head at the starting point of the saddle area is 10.50 m, which is 1.21 times of the designed head. The efficiency of the high-efficiency zone of the prefabricated pump station is 58.0~63.0%, and the corresponding flow range is 0.62Q_d~1.41Q_d (21.0~48.0 m^3/h). The uniformity of the inlet flow rate of impeller of pump 1 is 74.70%, and that of pump 2 is 75.57%. The flow fields of water pumps on both sides are inconsistent. The results of the flow field indicate that there are severe back flow phenomena at the prefabricated bucket intake, more back flow in the bucket, and many eddies on the side wall. With the increase in flow rate, the eddy structure at the intake expands continuously and moves towards the center area, which has a negative impact on the flow field in the center area. The research results of this paper can provide a theoretical reference for the research and operation of the same type of prefabricated pumping stations.

Keywords: prefabricated pumping station; centrifugal pump; energy characteristics; internal flow field; test

Citation: Xie, C.; Yuan, Z.; Feng, A.; Wang, Z.; Wu, L. Energy Characteristics and Internal Flow Field Analysis of Centrifugal Prefabricated Pumping Station with Two Pumps in Operation. *Water* **2022**, *14*, 2705. https://doi.org/10.3390/w14172705

Academic Editors: Changliang Ye, Xijie Song and Ran Tao

Received: 1 August 2022
Accepted: 29 August 2022
Published: 30 August 2022

Publisher's Note: MDPI stays neutral with regard to jurisdictional claims in published maps and institutional affiliations.

Copyright: © 2022 by the authors. Licensee MDPI, Basel, Switzerland. This article is an open access article distributed under the terms and conditions of the Creative Commons Attribution (CC BY) license (https://creativecommons.org/licenses/by/4.0/).

1. Introduction

With the worsening weather in recent years, some areas are often damaged by flood disasters, and the role of pumping stations is becoming increasingly prominent. Traditional pumping stations are mainly concrete pumping stations, which are costly, have a long construction period, and consume a lot of manpower and material resources. They cannot be moved after construction [1]. In this context, prefabricated pumping stations have gradually developed. Prefabricated pumping stations originated in Europe at the earliest. They are small and movable new drainage and irrigation equipment that combine all the components of the pumping stations into one unit. Compared with the traditional concrete pumping station, the prefabricated pumping station has the advantages of simple installation, short construction period, small area, good saving of land resources, and good economic benefits. A prefabricated pumping station is a power water conveyance device that integrates pumps, cylinders, pipes, and other components. At present, the research on prefabricated pumping stations is mostly focused on engineering applications. The flow pattern in the prefabricated pumping station is quite complex, and the bad flow pattern often affects the stable operation of the pumping station, causing cavitation and vibration of the pump and even damaging parts of the pump in serious cases. However, there is little research on the operation capacity and internal flow field of the prefabricated pumping station. The hydraulic performance and internal flow characteristics of the prefabricated pumping station are not clear, which makes it impossible to operate efficiently, stably, and safely. Therefore, it is necessary to study the prefabricated pumping station in depth.

Some scholars have carried out some research on the prefabricated pumping station; for example, related scholars pointed out that integrated prefabricated pumping station has great advantages in volume, service efficiency, etc. [2], analyzed the strength of integrated prefabricated pumping station cylinder [3], analyzed the sedimentation characteristics of solids when multiphase flow in integrated prefabricated pumping station [4,5], and analyzed the installation parameters of pumps in prefabricated pumping station [6]. There are few studies on the integrated pumping station. The research focuses on engineering application, structural design, and deposition of multiphase flow solids. There is little research on its hydraulic performance and internal flow characteristics. Studies have shown the advantages of using CFD methods for centrifugal pump characteristics and flow field studies; for example, related scholars have used CFD methods to analyze the energy characteristics of centrifugal pumps [7,8] and the pressure, velocity, and streamline the distribution of centrifugal pumps [9], and the literature [10] shows that the energy characteristics of centrifugal pumps can be studied using the SST k-ω turbulence model. Related scholars have combined numerical calculations with experiments to analyze and discuss the pump performance and internal flow patterns; for example, some scholars have compared the numerical calculation results with experimental data for multiple working conditions [11], studied the effect of cutouts on the guide vane on the pump performance and internal flow [12], revealed the performance and internal flow characteristics of multi-stage single suction centrifugal pumps, analyzed in detail the internal flow and pressure field of centrifugal pumps [13], and analyzed the effect of impeller vane number and angle on the performance of centrifugal pumps [14].

In this paper, considering the influence of structural components on the internal flow field, the local dimensions of structural components are not simplified, and numerical calculations and experimental analysis are used for the study with a view to gaining insight into the hydraulic performance and internal flow characteristics of prefabricated pumping stations in actual operation, revealing the undesirable flow patterns inside prefabricated pumping stations, and providing theoretical guidance for the optimal design and operation of prefabricated pumping stations. The research in this paper has certain academic significance and engineering application value.

2. Three-Dimensional Modeling and Numerical Calculation Setup

2.1. Calculation Model

In this paper, Solidworks software [15] is used to build the three-dimensional model of the whole pumping station. When modeling, the influence of the structural components such as motor, pipe, and flange in the pumping station on the water flow is considered. The overall height of the cylinder is $L = 1$ m, the diameter of the cylinder is $D = 1$ m, and the diameter of the intake and outlet is $R = 100$ mm. The pump used in this paper is a submerged centrifugal pump with a diameter of impeller $d = 100$ mm, number of vanes of 3 pieces, and speed $n = 2900$ r/min. The three-dimensional model is as shown in the Figure 1.

2.2. Meshing

In this paper, the water body parts inside the structural components are extracted by SolidWorks software, and the three-dimensional model is meshed in Mesh software. The mesh calculation area mainly includes the inlet section, prefabricated barrel, impeller, guide vane, and outlet section. Because the fluid excises more structural components and is more complex, this paper uses non-structural tetrahedral mesh for mesh division, and the mesh is better adapted [16–18]. The fluid calculation mesh of the centrifugal prefabricated pumping station is shown in Figure 2.

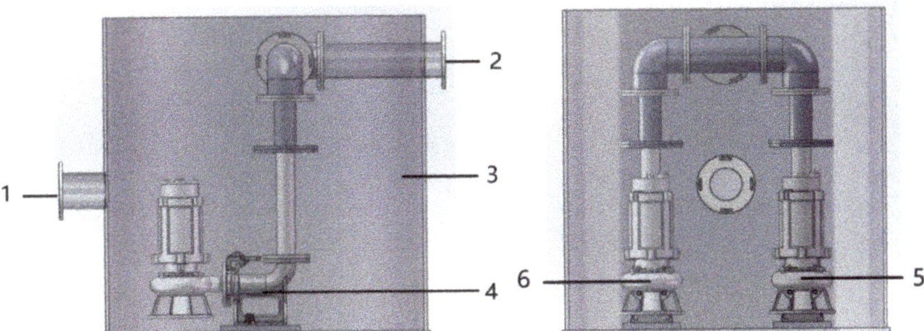

Figure 1. General assembly drawing of three-dimensional model of centrifugal prefabricated pumping station. 1. Inlet; 2. outlet; 3. round prefabricated barrels; 4. couplers; 5. submersible centrifugal pump 1; 6. submersible centrifugal pump 2.

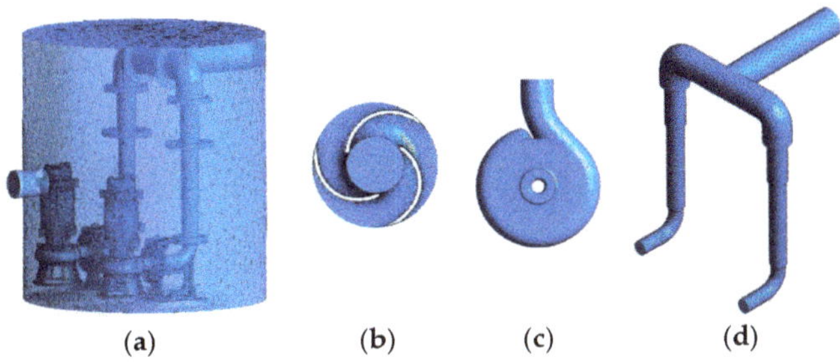

Figure 2. Grid of each calculation domain. (**a**) Prefabricated barrel grid; (**b**) impeller grid; (**c**) worm gear grid; (**d**) outlet section grid.

In this paper, seven scenarios with different numbers of meshes under the design condition (Q_d = 33.93 m^3/h) are selected for the numerical calculation of the centrifugal prefabricated pumping station, and the efficiency of the centrifugal prefabricated pumping station is used as the evaluation index. It can be seen from Figure 3 that when the number of grids is between 0.9 million and 3 million, the efficiency changes greatly and is unstable. When the number of grids reaches 3.2 million, the efficiency curve basically remains unchanged, indicating that the increase in the number of grids has little impact on the calculation results [19]. Considering the computer performance and the accuracy of the calculation results, this paper selects 3.2 million grids for numerical calculation.

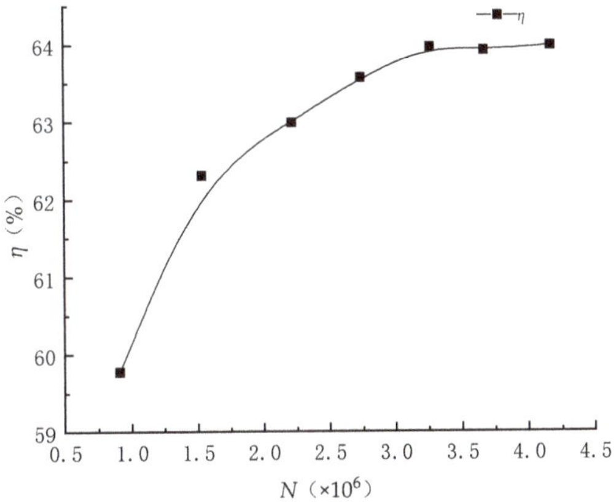

Figure 3. Grid irrelevance analysis diagram.

2.3. Boundary Conditions and Turbulence Model

The inlet of a centrifugal prefabricated pumping station is on the side of the prefabricated barrel, on the same level as the outlet. The condition for the inlet is set to Total Pressure, and the pressure magnitude is set to one atmosphere. The outlet is the outlet of the centrifugal prefabricated pumping station, and the outlet condition is set to Normal Speed. The solid wall is set to a no-slip boundary condition [20], the fluid has zero velocity near the wall, and the effect of wall roughness is not taken into account. The surface wall condition of the impeller is also set to the Rotating wall, and the surface wall condition of the cylinder, worm casing, and outlet section is set to the Static wall, applying the boundary no-slip boundary condition. In this paper, the two pumps are arranged symmetrically on the central axis of the cylinder, two rotation axes are set for the numerical calculation, the vertical direction of the impeller centers of pump 1 and pump 2 are used to determine the respective rotation axes, the coordinate positions of pump 1 rotation axis are (0.235, 0.1, 0.265), and pump 2 rotation axis coordinates are (−0.235, 0.1, −0.265).

In the calculation area, the calculation area of the inlet section, the cylinder, the worm housing, and the outlet section are set to the static domain, and the calculation area of the impeller is set to the rotational domain. In this paper, the Frozen Rotor interfacing model is used as the dynamic and static interfacing model for the fluid connection between the impeller and the worm gear. The numerical calculation of the centrifugal prefabricated pumping station uses the Reynolds time-averaged N-S equation, and the turbulence model uses the SST k-ω [21,22] turbulence model, which absorbs the advantages of the standard k-ε model and the standard k-ω model, and adopts the automatic function at the boundary layer, which can better capture the flow at the boundary layer. The diffusion term and pressure gradient are represented by the finite volume method based on finite elements, and the convective term is in the High-Resolution Scheme. In the calculation, the pressure of the flow field is P; the velocities in the x, y, and z directions are u, v, and w; the convergence conditions of the turbulent kinetic energy k equation and the dissipation rate ε are set to 10^{-5}, and in principle, the smaller the residuals are, the better.

2.4. Calculation Formula

2.4.1. Control Equations

Turbulence control equation (N-S equation):

$$\frac{\partial(\rho u_i)}{\partial t}+\frac{\partial(\rho u_i u_j)}{\partial x_j}=-\frac{\partial P}{\partial x_i}+[\mu(\frac{\partial u_i}{\partial x_j}+\frac{\partial u_j}{\partial x_i})]+F_i \tag{1}$$

where t is time (s); ρ is fluid density (kg/m^3); x_i and x_j are spatial coordinates; u_i and u_j are the velocity components of the fluid parallel to the corresponding axes x_i and x_j, respectively, and F_i is the volume force component in the i-direction; μ is the fluid dynamic viscosity coefficient; P is the pressure (Pa).

The transport equation of the SST k-ω turbulence model can be expressed as:

$$\frac{\partial(\rho k)}{\partial t}+\frac{\partial(\rho k u_i)}{\partial x_i}=\frac{\partial}{\partial x_j}[(\mu+\frac{\mu_t}{\sigma_k})\frac{\partial k}{\partial x_j}]+G_k-Y_k+S_k \tag{2}$$

$$\frac{\partial(\rho \omega)}{\partial t}+\frac{\partial(\rho \omega u_i)}{\partial x_i}=\frac{\partial}{\partial x_j}[(\mu+\frac{\mu_t}{\sigma_\omega})\frac{\partial \omega}{\partial x_j}]+G_\omega-Y_\omega+S_\omega+D_\omega \tag{3}$$

where G_k, G_ω is the generating term of the equation; Y_k, Y_ω is the generating term of the diffusive action; S_k, S_ω is the user-defined source term; D_ω is the term generated by the orthogonal divergence; k is the turbulent kinetic energy; ω is the turbulent special dissipation; μ_t is turbulent dynamic viscosity coefficient.

2.4.2. Hydraulic Performance Prediction

Centrifugal prefabricated pumping station head, expressed by the following equation [23,24]:

$$H_{net} = \left(\frac{\int\limits_{s2} P_2 u_t ds}{\rho Q g} + H_2 + \frac{\int\limits_{s2} u_2^2 u_{t2} ds}{2Qg} \right) - \left(\frac{\int\limits_{s1} P_1 u_t ds}{\rho Q g} + H_1 + \frac{\int\limits_{s1} u_1^2 u_{t1} ds}{2Qg} \right) \quad (4)$$

where the first term on the right side of the equation is the total pressure at the outlet section of the prefabricated barrel, and the second term is the total pressure at the inlet section of the prefabricated barrel. Q—flow rate, m^3/s; H_1, H_2—prefabricated barrel inlet and outlet section elevation, m.

s_1, s_2—prefabricated barrel inlet and outlet section area; u_1, u_2—prefabricated barrel inlet and outlet flow velocity at each point, m/s; u_{t1}, u_{t2}—prefabricated barrel inlet and outlet section flow velocity normal component at each point, m/s.

P_1, P_2—prefabricated barrel inlet and outlet section at each point of the static pressure, Pa; g—gravitational acceleration, m/s^2.

The efficiency of centrifugal prefabricated pumping stations is [25–27]:

$$\eta = \frac{\rho g Q H_{net}}{N_1 + N_2} \quad (5)$$

where N_1—shaft power of pump 1, N_2—shaft power of pump 2.

The shaft power of the centrifugal prefabricated pumping station is [28,29]:

$$N = \frac{\pi}{30} T n \quad (6)$$

where T—torque, N—m; n—rotational speed, r/min.

2.4.3. Uniformity of Flow Velocity Distribution

The uniformity of axial velocity distribution V_{zu} of the section at the impeller inlet reflects the water inlet quality of the impeller, and the closer V_{zu} is to 100%, the more uniform the water inlet of the impeller is, and its calculation formula is as follows [30]:

$$V_{zu} = \left\{ 1 - \frac{1}{\overline{v_a}} \sqrt{\left[\sum_{i=1}^{n} (v_{ai} - \overline{v_a})^2 \right] / n} \right\} \times 100\% \quad (7)$$

where V_{zu}—uniformity of flow velocity distribution at impeller inlet, %; V_a—arithmetic mean of axial flow velocity at impeller inlet; V_{ai}—axial velocity of each calculation unit at impeller inlet, m/s; n—number of calculation units at impeller inlet.

3. Energy and Internal Flow Characteristics Analysis

In this paper, the numerical calculation results under different flow conditions are extracted to analyze the flow field characteristics inside the centrifugal prefabricated pumping station. The numerical calculations are divided into 12 flow conditions, which are 11.31 m^3/h (0.33Q_d), 16.96 m^3/h (0.50Q_d), 22.62 m^3/h (0.67Q_d), 28.27 m^3/h (0.83Q_d), 33.93 m^3/h (1.00Q_d), 39.58 m^3/h (1.17Q_d), 45.24 m^3/h (1.33Q_d), 50.89 m^3/h (1.50Q_d), 56.55 m^3/h (1.67Q_d), 62.20 m^3/h (1.83Q_d), 67.86 m^3/h (2.00Q_d), 73.51 m^3/h (2.17Q_d), and 79.17 m^3/h (2.33Q_d); design flow working condition is Q_d = 33.93 m^3/h.

3.1. *Energy Characteristics Analysis*

The numerical calculation results of the centrifugal prefabricated pumping station at different flow rates are extracted, and the head and efficiency of centrifugal prefabricated pumping station are calculated by Equations (4)–(6), followed by drawing the energy characteristic curve of the centrifugal prefabricated pumping station. The energy characteristics are shown in Table 1 and Figure 4.

Table 1. Energy characteristics data table.

Flow Rate Q (m^3/h)	Head H (m)	Efficiency η (%)
11.31 (0.33Q_d)	10.50	45.05
16.96 (0.50Q_d)	10.33	55.28
22.62 (0.67Q_d)	9.91	59.77
28.27 (0.83Q_d)	9.39	63.22
33.93 (1.00Q_d)	8.66	63.96
39.58 (1.17Q_d)	7.77	61.78
45.24 (1.33Q_d)	6.84	60.36
50.89 (1.50Q_d)	5.91	57.54
56.55 (1.67Q_d)	4.93	53.17
62.20 (1.83Q_d)	3.94	47.15
67.86 (2.00Q_d)	2.92	39.18
73.51 (2.17Q_d)	1.90	28.97
79.17 (2.33Q_d)	0.57	9.98

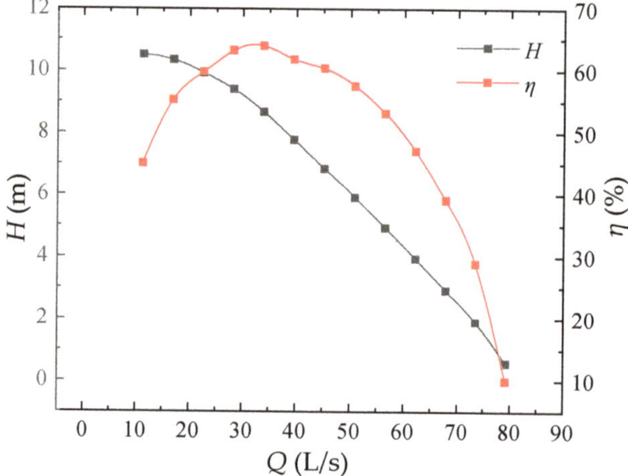

Figure 4. Energy characteristic curve at different flow rates.

The calculation results of the energy characteristics of the centrifugal prefabricated pumping station show that the efficiency of the prefabricated pumping station increases with the flow rate under the small flow rate condition (Q = 11.31~33.93 m^3/h), and at the flow condition 0.33Q_d (Q = 11.31 m^3/h), the centrifugal prefabricated pumping station enters the flow instability condition (saddle zone effect area), and this phenomenon can be obviously found through the head, the head increases less from 0.50Q_d to 0.33Q_d, and the slope of growth here is close to 0 from the head curve and reaches the maximum value at the design condition (Q_d = 33.93 m^3/h) with the maximum efficiency value of 63.96% and the head is 8.66 m. The head at the starting point of the saddle area is 10.50 m, which is 1.21 times the design head, indicating that the operable head range for small flow is small and the operable head range for large flow is wide. Under the high flow rate condition (Q = 33.93~79.17 m^3/h), the efficiency decreases gradually with the increase in flow rate. The slope of the efficiency curve change increases with increasing flow rate. When the flow rate is between 21.0~48.0 m^3/h, the prefabricated pumping station is in the high-efficiency zone (the high-efficiency zone is defined as the range of flows where the efficiency of the optimal efficiency point decreases by 5%); at this time, the pumping station device efficiency is around 58.0~63.0%. The head curve of the prefabricated pumping station gradually decreases with the increase in flow, from 10.50 m to 0.57 m, and the efficiency curve is parabolic with the increase in flow; the head curve is approximately straight with a small change in slope.

Based on the post-processing of the numerical calculation results, the uniformity of flow velocity at the impeller inlet of pump 1 and pump 2 is calculated by Equation (7). The uniformity of flow rate is shown in Table 2 and Figure 5.

Table 2. Uniformity of impeller inlet flow rate at different flow rates.

Flow Rate Q (m^3/h)	Water Pump 1 Impeller Inlet Flow Rate Uniformity (%)	Water Pump 2 Impeller Inlet Flow Rate Uniformity (%)
11.31 (0.33Q_d)	68.55	66.05
16.96 (0.50Q_d)	70.55	73.03
22.62 (0.67Q_d)	73.93	74.74
28.27 (0.83Q_d)	74.59	75.40
33.93 (1.00Q_d)	74.70	75.57
39.58 (1.17Q_d)	74.61	75.52
45.24 (1.33Q_d)	74.65	75.55
50.89 (1.50Q_d)	74.70	75.61
56.55 (1.67Q_d)	74.74	75.66
62.20 (1.83Q_d)	74.77	75.70
67.86 (2.00Q_d)	74.79	75.72
73.51 (2.17Q_d)	74.79	75.70
79.17 (2.33Q_d)	74.78	75.67

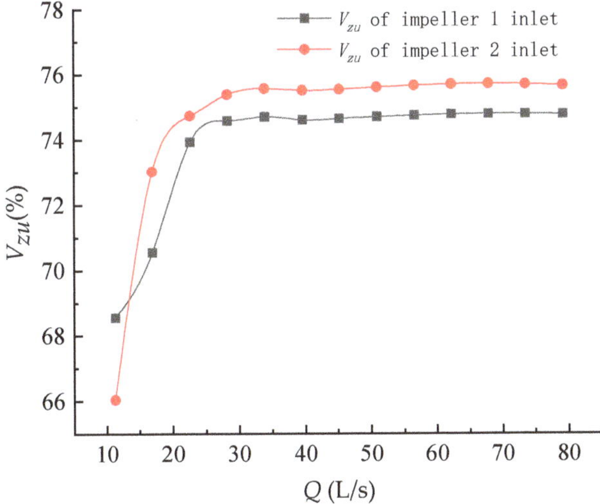

Figure 5. Uniformity of impeller inlet flow rate under different working conditions.

As can be seen from Figure 5, the variation pattern of flow uniformity with increasing flow rate at the impeller inlet of pump 1 and pump 2 is essentially the same. Under the small flow rate condition, the uniformity of flow velocity at the impeller inlet increases with the increase in flow rate, and when the flow rate reaches 0.83Q_d (Q = 28.27 m^3/h), the uniformity of flow velocity at the impeller inlet remains basically the same with the increase in flow rate. Overall, the uniformity of flow velocity at the impeller inlet of pump 2 is greater than that of pump 1; at the design working condition (Q_d = 33.93 m^3/h), the uniformity of flow velocity at the impeller inlet of pump 1 is 74.70%, and that of pump 2 is 75.57%, with a difference of 0.87%, which is caused by the different uniformity of flow velocity at the impellers of pump 1 and pump 2 due to bias flow. When the flow condition is less than $0.83Q_d$ (Q = 28.27 m^3/h), the uniformity of flow velocity at the impeller inlet decreases significantly, which indicates that the unevenness of the flow pattern inside the prefabricated barrel of the centrifugal prefabricated pumping station increases at this time and cannot provide a better impeller inlet water flow pattern.

3.2. Analysis of Internal Flow Characteristics

In the numerical calculation results of the centrifugal prefabricated pumping station, six working conditions of $0.33Q_d$, $0.67Q_d$, $1.00Q_d$, $1.33Q_d$, $1.67Q_d$, and $2.00Q_d$ are selected for the analysis of the flow field inside the prefabricated pumping station, and in order to better depict the flow field inside the prefabricated pumping station, four characteristic sections as shown in Figure 6 are selected for analysis in this paper. The A1 section is the horizontal cross section at the impeller inlet of the submersible centrifugal pump, A2 is the cross section at the center of the precast barrel inlet, A3 is the horizontal cross section at the highest liquid level of the precast barrel, and A4 is the vertical cross section at the center of the precast barrel.

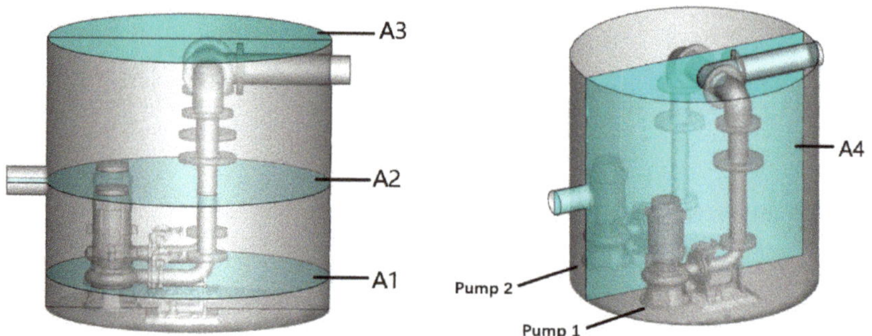

Figure 6. Schematic diagram of centrifugal prefabricated pumping station cross-section.

The velocity and streamline distributions of A1–A4 sections of centrifugal prefabricated pumping stations at $0.33Q_d$, $0.67Q_d$, $1.00Q_d$, $1.33Q_d$, $1.67Q_d$, and $2.00Q_d$ flow conditions are shown below.

Cross-section A1 flow velocity and streamline distribution is shown in Figure 7. It can be seen from Figure 7 that in the small flow conditions, the prefabricated barrel inlet side flow velocity is less than the outlet side, the impeller inlet flow velocity distribution is also uneven, the flow velocity alternates between multiple velocity classes at the impeller inlet, and there is a flow stratification effect, pump 1 and pump 2 in the middle of the water flow streamlines parallel to each other, in the prefabricated barrel along the water flow direction to see the left and right sides of the wall have vortex, and near the pump 2 vortex area is larger than near the pump 1, the coupler waterward surface on both sides of the vortex also exists, the vortex at the wall is mainly caused by the backflow of the water impacting to the back wall of the prefabricated barrel, and the backflow at the coupler is caused by the local structural features blocking the water. Under the high flow condition, the flow velocity at the impeller inlet is distributed periodically, the flow velocity distribution in the prefabricated barrel is more dispersed, the streamline is disorderly, the vortex area on both sides of the prefabricated barrel wall increases compared with the low flow condition, and the area of the vortex area for flow conditions is about twice as large as for small flow conditions.

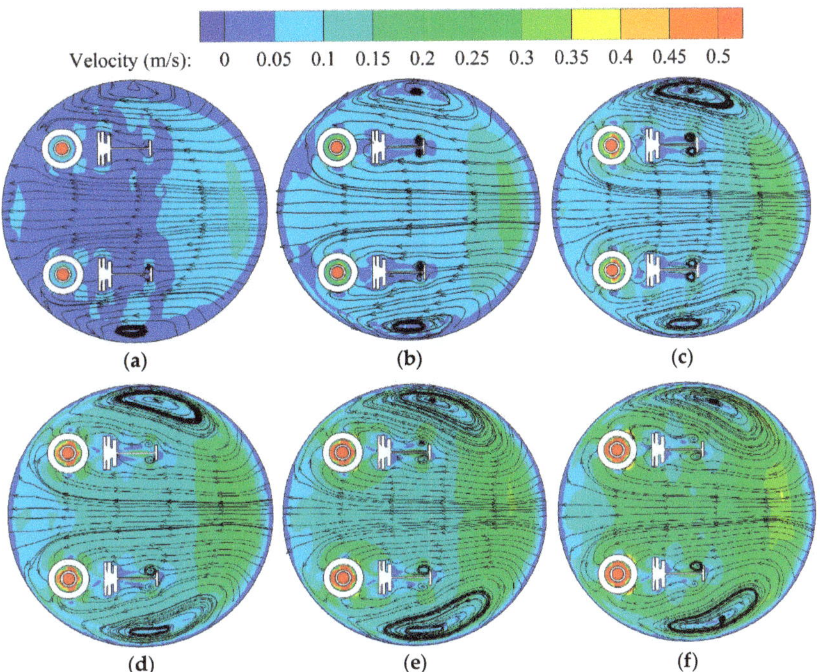

Figure 7. Cross-section A1 flow velocity and streamline distribution. (**a**) $0.33Q_d$, (**b**) $0.67Q_d$, (**c**) $1.00Q_d$, (**d**) $1.33Q_d$, (**e**) $1.67Q_d$, and (**f**) $2.00Q_d$.

Cross-section A2 flow velocity and streamline distribution is shown in Figure 8. It can be obtained from Figure 8 that the inlet is a high-speed water flow area, which is fan-shaped diffusion near the inlet of the prefabricated barrel, and the flow velocity decreases in a gradient, with the increase in the flow rate, the speed at the inlet is accelerated, and the fan-shaped diffusion area becomes larger. There are vortices on both sides of the inlet and both sides of the barrel wall, and the vortices on both sides of the inlet gradually move to the center of the prefabricated barrel with the increase in flow, and the streamlines in the center of the prefabricated barrel are parallel, and there are streamlines intersecting at pump 1 and pump 2, and the distribution of streamlines is more disorderly. Analyzed from the change in working conditions, the vortices at the inlet and sidewall both increased with the increase in flow rate. Analyzed from the position, the vortex on both sides of the inlet has basically the same area compared to each other, indicating that the bias flow does not affect the upstream flow pattern, so the symmetry is better. The area of vortex area near pump 2 is larger than that near pump 1, indicating that there is an obvious bias flow inside the prefabricated barrel.

Cross-section A3 flow velocity and streamline distribution is shown in Figure 9. It can be obtained from Figure 9 that there is a low-velocity zone at the side wall of the precast barrel wall, which is mainly due to the side wall effect, and there are three low-velocity zones in the center of the precast barrel, which is mainly due to the existence of flanges in these three places (as shown in Figure 6). There are vortices on both walls of the precast barrel, the area of the vortex region is small compared to the A1 and A2 cross sections. Analyzed in terms of the change in working conditions, and the area of the vortex zone increases with the increase in the flow rate. With the increase in flow rate, the velocity distribution in the prefabricated barrel changed obviously, producing a crescent-shaped high-speed zone on the inlet side and multiple irregular high-speed zones on the outlet side, and the streamline in the prefabricated barrel was disordered. When analyzed in terms of location, the area of the vortex zone near pump 2 is still larger than that near pump 1, which is also caused by the bias flow effect in the prefabricated barrel.

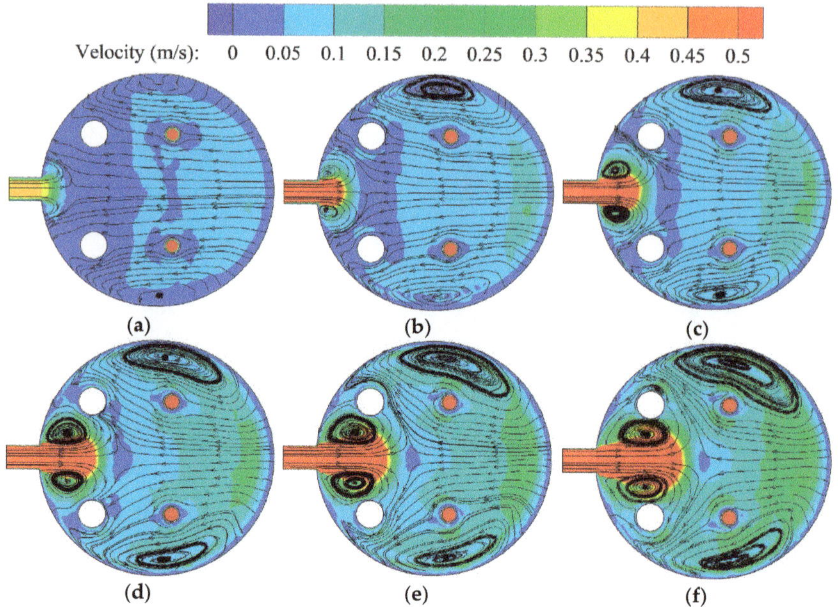

Figure 8. Cross-section A2 flow velocity and streamline distribution. (**a**) $0.33Q_d$, (**b**) $0.67Q_d$, (**c**) $1.00Q_d$, (**d**) $1.33Q_d$, (**e**) $1.67Q_d$, and (**f**) $2.00Q_d$.

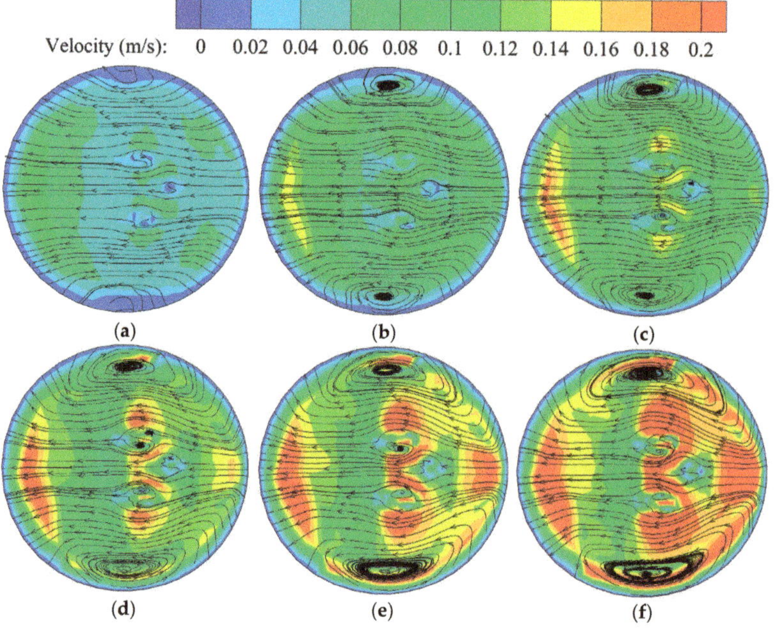

Figure 9. Cross-section A3 velocity and streamline distribution. (**a**) $0.33Q_d$, (**b**) $0.67Q_d$, (**c**) $1.00Q_d$, (**d**) $1.33Q_d$, (**e**) $1.67Q_d$, and (**f**) $2.00Q_d$.

Cross-section A4 flow velocity and streamline distribution is shown in Figure 10. It can be obtained from Figure 10 that the water flow in the inlet and outlet pipes is a high-speed zone, and under the low flow condition, there are vortices on the upper and lower sides of the inlet, and there are many small vortices at the upper and lower sides of the outlet. Under the high flow condition, the velocity at the water inlet decreases in a gradient toward the center of the prefabricated barrel, and the vortex at the water inlet moves toward the center. The vortex on the upper side of the prefabricated barrel inlet is gradually smaller than the vortex on the lower side, and the velocity and streamline in the prefabricated barrel are more chaotic. From the analysis of the change in working conditions, with the increase in flow rate, the velocity of water at the inlet is accelerated, and the vortex structure at the inlet becomes larger and pushes it to move to the center continuously, which affects the flow field at the center of the prefabricated barrel. Analysis from the position of, due to the influence of the flange structure, the flow of water above the outlet pipe is disturbed, and the vortex is generated.

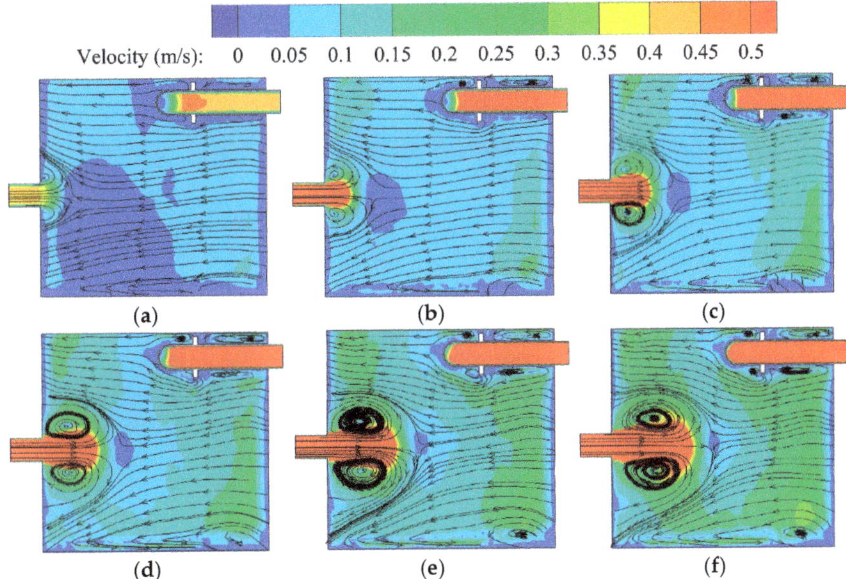

Figure 10. Cross-section A4 velocity and streamline distribution. (**a**) $0.33Q_d$, (**b**) $0.67Q_d$, (**c**) $1.00Q_d$, (**d**) $1.33Q_d$, (**e**) $1.67Q_d$, and (**f**) $2.00Q_d$.

Prefabricated pumping station wall pressure distribution is shown in Figure 11. From Figure 11, the pressure on the inlet side of the prefabricated barrel of the prefabricated pumping station gradually decreases with the increase in the flow rate until the maximum flow rate of $2.00Q_d$; only a small part of the high-pressure area exists on the upper side of the inlet and the upper side of the prefabricated barrel. Prefabricated pumping station prefabricated barrel outlet side pressure with the flow rate increases, the low-pressure area gradually decreases; until the maximum flow rate of $2.00Q_d$, the low-pressure area basically does not exist. In the centrifugal pump outlet connection pipeline, with the increase in flow, the low-pressure area becomes larger, mostly concentrated in the elbow of the double pump sink pipe. As the flow rate increases, the low-pressure area also appears on the waterward side of the centrifugal pump motor. Prefabricated barrel pressure distribution is not uniform, mainly in the inlet and outlet side is not consistent, along the inlet and outlet water axis, the symmetry of the left and right sides is slightly better, but also not completely symmetrical, can also be found in the phenomenon of partial flow.

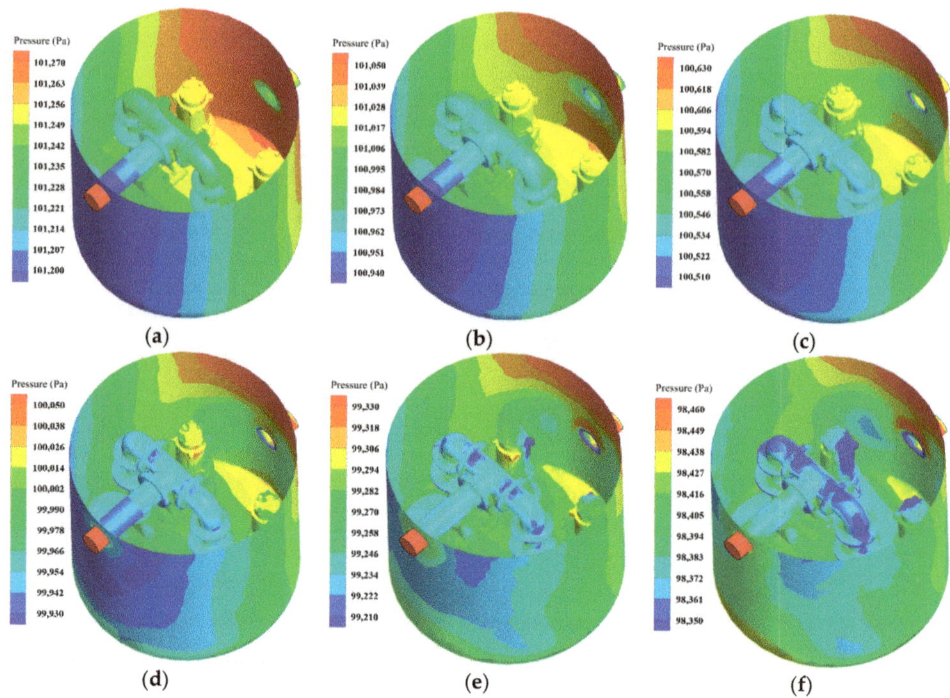

Figure 11. Prefabricated pumping station wall pressure distribution diagram. (**a**) $0.33Q_d$, (**b**) $0.67Q_d$, (**c**) $1.00Q_d$, (**d**) $1.33Q_d$, (**e**) $1.67Q_d$, and (**f**) $2.00Q_d$.

4. Experiment Equipment, Test and Result Analysis

4.1. Test Bench Introduction

Centrifugal prefabricated pumping station test bench has the following parts: Round prefabricated barrel, Back-flow tank, Submersible centrifugal pump, Coupler, Inlet and Outlet pipes, Electromagnetic Flow meter, Pipeline pump, and PLC frequency control cabinet. The total length of the test bench is about 5 m, the diameter of the pipe is 100 mm, and the whole is a circulating system. The whole test bench is made of acrylic material to achieve transparent visualization and to be able to clearly observe the flow pattern of water inside the centrifugal pumping station. Figure 12 shows the sketch of the centrifugal prefabricated pumping station test bench, Figure 13 shows the three-dimensional model of the centrifugal prefabricated pumping station test bench, and Figure 14 shows the physical drawing of the centrifugal prefabricated pumping station test bench. The flow rate is measured by an electromagnetic flowmeter (ZEF-DN100, range 0~120 m^3/h, accuracy $\pm 0.5\%$), and the flow pattern is captured by a high-speed camera (OLYMPUS i-SPEED 3, working range 2000 fps full resolution, accuracy ± 1 μs).

In this test, the centrifugal pump 3 was first adjusted to the rated speed $n = 2900$ r/min, and then the flow rate of the inlet of the prefabricated pumping station was adjusted to the design flow rate ($Q_d = 33.93$ m^3/h) by controlling the pipeline pump 6, and then the high-speed camera was used to take pictures of the internal flow state of the integrated, prefabricated pumping station.

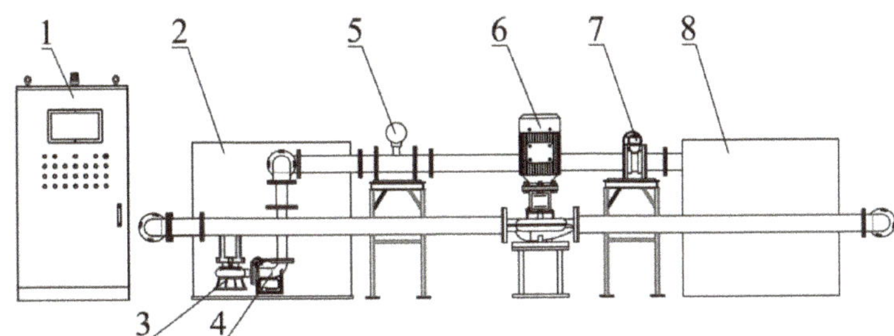

Figure 12. Sketch of centrifugal prefabricated pumping station test bench. 1. PLC variable frequency control cabinet; 2. round prefabricated barrels; 3. submersible centrifugal pump; 4. couplers; 5. electromagnetic flowmeter; 6. pipe pump; 7. turbo butterfly valve; 8. backflow tank.

Figure 13. Three-dimensional model of centrifugal prefabricated pumping station.

Figure 14. Physical drawing of centrifugal prefabricated pumping station test bench.

4.2. Analysis of Experimental Results

In this paper, a high-speed camera is used to photograph the internal flow pattern of the prefabricated barrel under the design flow condition (Q_d = 33.93 m^3/h), and a tracer red line is used to show the water flow to obtain pictures of the internal flow of the centrifugal prefabricated pumping station at different moments under different orientations.

Flow pattern of prefabricated pumping station. is shown in Figure 15. From Figure 15, it can be seen that the tracer line on both sides of the barrel wall oscillates with the flow of water. The direction of oscillation is from the outlet to the inlet, which is consistent with the direction of streamline in the numerical calculation results. At the intake, the tracer line can be seen to swirl with the water flow on both sides, which indicates that there is a backflow at the intake, and a vortex is generated. The internal flow characteristics are similar to those of the numerical calculation; the experimental study of flow pattern visualization is a verification of the numerical calculation results. The large variation in the position of the tracer red line under different moments and the large oscillation also indicate the turbulence of the flow pattern in the prefabricated barrel of the centrifugal prefabricated pumping station and the inconsistency of the flow line.

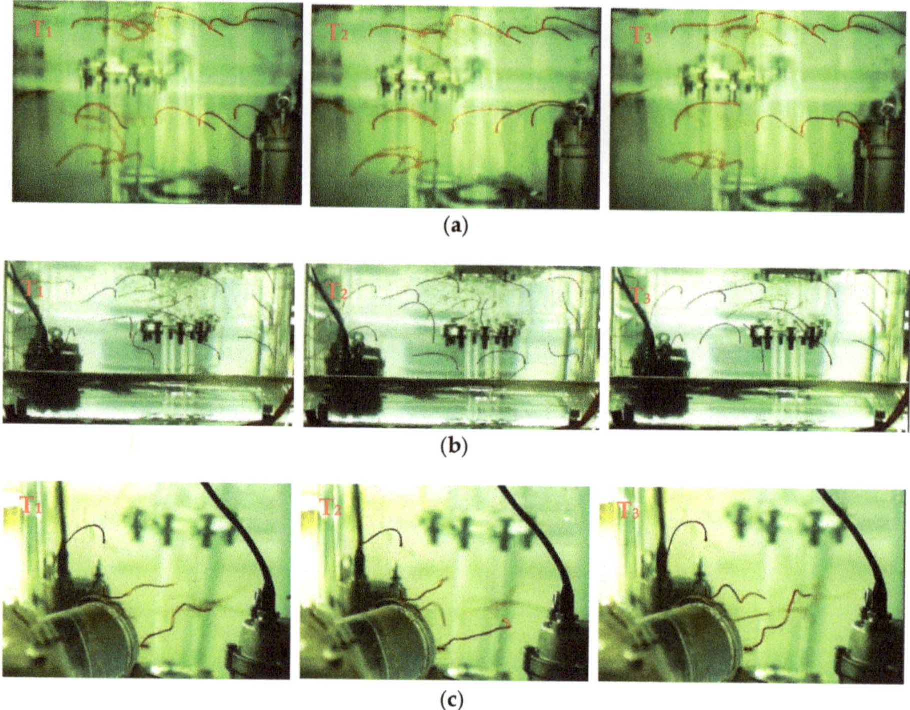

Figure 15. Flow pattern of prefabricated pumping station. (**a**) Left side of prefabricated barrel; (**b**) right side of prefabricated barrel; (**c**) filming at the inlet of the prefabricated barrel.

5. Conclusions

This paper takes centrifugal prefabricated pumping station as the research object; numerical calculation of different working conditions under double pumping operation conditions and analysis and discussion of internal flow characteristics of prefabricated pumping station are carried out. An acrylic visualization prefabricated pumping station test bench was built to verify the numerical calculation results of centrifugal prefabricated pumping station by experimentally filming the internal flow pattern of the prefabricated pumping station, and the main conclusions are as follows.

(1) The calculation shows that the maximum value is reached at the design working condition (Q_d = 33.93 m^3/h); the maximum efficiency value is 63.96%; the head is 8.66 m; the head at the starting point of the saddle area is 10.50 m, which is 1.21 times of the design head; the operable head range for small flow is small; and the operable head range for large flow is wide. The efficiency of the prefabricated pumping station in the high-efficiency zone is 58.0~63.0%, corresponding to the flow range of 0.62Q_d~1.41Q_d (21.0~48.0 m^3/h).

(2) Through the numerical calculation, it can be obtained that the impeller inlet flow uniformity increases with the increase in flow rate, and it is basically maintained at the same value from 0.83Q_d to 2.33Q_d. The impeller inlet flow uniformity of pump 2 is greater than that of pump 1. At the design working condition (Q_d = 33.93 m^3/h), the impeller inlet flow uniformity of pump 1 is 74.70%, and that of pump 2 is 75.57%. The inlet flow fields of the pumps on both sides are not consistent. When the flow condition is less than 0.83Q_d (Q = 28.27 m^3/h), the uniformity of flow velocity at the impeller inlet increases significantly and cannot provide a better impeller inlet flow pattern.

(3) Through the numerical calculation of velocity and pressure cloud diagrams as well as the experimental flow analysis, it can be obtained that the centrifugal prefabricated pumping station has a serious backflow phenomenon at the inlet, and multiple vortexes are generated. As the flow rate increases, the vortex structure at the inlet expands and moves to the central area, which has a negative impact on the flow field in the central area. The flow pattern in the prefabricated barrel was unstable, and there was a partial flow on the left and right sides.

6. Suggestions

This paper reveals the energy characteristics and internal flow field of centrifugal prefabricated pumping stations under double pump operation conditions through numerical calculation and experimental analysis. However, because of the limited page, the research work has achieved certain results, but there are still many problems that can be studied and need to be further expanded and deepened in future work. In the future, we will start the multi-disciplinary optimization design of centrifugal prefabricated pumping station with multiple working conditions, and the research results of this paper can provide the basis for the optimization design of centrifugal prefabricated pumping station.

Author Contributions: Concept design, C.X. and Z.Y.; Numerical calculation, Z.Y., A.F. and. Z.W.; Experiment and data analysis, Z.Y., A.F., Z.W. and L.W.; Manuscript writing, C.X. and Z.Y. All authors have read and agreed to the published version of the manuscript.

Funding: This research was funded by Anhui Province Natural Science Funds for Youth Fund Project, grant number 2108085QE220. Key scientific research project of Universities in Anhui Province, grant number KJ2020A0103. Anhui Province Postdoctoral Researchers' Funding for Scientific Research Activities, grant number 2021B552. Anhui Agricultural University President's Fund, grant number 2019zd10. Stabilization and Introduction of Talents in Anhui Agricultural University Research Grant Program, grant number rc412008.

Institutional Review Board Statement: Not applicable.

Informed Consent Statement: Not applicable.

Data Availability Statement: Not applicable.

Conflicts of Interest: The authors declare no conflict of interest.

References

1. Wang, D. Research on design development and application of integrated prefabricated pumping station. *Gen. Mach.* **2014**, *7*, 87–88.
2. Fang, A.B. Research and application of integrated drainage pumping station. *Build. Mater. Decor.* **2020**, *1*, 218–219.
3. Zhang, Z.; Wang, K.; Chen, K.; Yang, J.; Wang, S.; Wang, Y. Effects of different operation modes on flow characteristics and cylinder strength of integrated prefabricated pumping stations. *China Rural. Water Conserv. Hydropower* **2019**, *4*, 162–167.
4. Li, Q.; Kang, C.; Teng, S.; Li, M. Optimization of Tank Bottom Shape for Improving the Anti. *Deposition Performance of a Prefabricated Pumping Station. Water* **2019**, *11*, 602.
5. Wang, K.; Hu, J.; Liu, H.; Zhang, Z.; Zou, L.; Lu, Z. Research on the Deposition Characteristics of Integrated Prefabricated Pumping Station. *Symmetry* **2020**, *12*, 760. [CrossRef]
6. Zhang, B.; Cheng, L.; Xu, C.; Wang, M. The Influence of Geometric Parameters of Pump Installation on the Hydraulic Performance of a Prefabricated Pumping Station. *Energies* **2021**, *14*, 1039. [CrossRef]
7. Iacob, N.; Drăgan, N. Dynamic Analysis of a Centrifugal Pump using CFD and FEM Methods. *Hidraulica* **2019**, *4*, 29–37.
8. Ramakrishna, R.; Hemalatha, S.; Rao, D.S. Analysis and performance of centrifugal pump impeller. *Mater. Today Proc.* **2022**, *50*, 5. [CrossRef]
9. Kumar, S.V.; Han, X.; Kang, Y.; Li, D.; Zhao, W.; Selamat, F.E.; Wan Izhan, W.H.I.; Baharudin, B.S.; Bellary, S.A.I. Upgradation in efficiency of centrifugal pump. *Asian J. Multidimens. Res.* **2021**, *10*, 32–34.
10. Tong, Z.; Xin, J.; Tong, S.; Yang, Z.; Zhao, J.; Mao, J. Internal flow structure, fault detection, and performance optimization of centrifugal pumps. *J. Zhejiang Univ.-Sci. A Appl. Phys. Eng.* **2020**, *21*, 85–117. [CrossRef]
11. Al-Obaidi, A.R. Numerical investigation on effect of various pump rotational speeds on performance of centrifugal pump based on CFD analysis technique. *Int. J. Modeling Simul. Sci. Comput.* **2021**, *12*, 2150045. [CrossRef]
12. Shunya, T.; Shinichi, K.; Shinichiro, E.; Masahiro, M. Effect of diffuser vane slit on rotating stall behavior and pump performance in a centrifugal pump. *J. Phys. Conf. Ser.* **2022**, *2217*, 012054.

13. Tan, M.; Lu, Y.; Wu, X.; Liu, H.; Tian, X. Investigation on performance of a centrifugal pump with multi-malfunction. *J. Low Freq. Noise Vib. Act. Control* **2020**, *40*, 740–752. [CrossRef]
14. Susilo, S.H.; Setiawan, A. Analysis of the number and angle of the impeller blade to the performance of centrifugal pump. *EUREKA Phys. Eng.* **2021**, *5*, 62–68. [CrossRef]
15. Ding, H.; Li, Z.; Gong, X.; Li, M. The influence of blade outlet angle on the performance of centrifugal pump with high specific speed. *Vacuum* **2019**, *159*, 239–246. [CrossRef]
16. Lila, A.; Mathieu, S.; Idir, B.; Smaine, K. Numerical Assessment of the Hydrodynamic Behavior of a Volute Centrifugal Pump Handling Emulsion. *Entropy* **2022**, *24*, 221. [CrossRef]
17. Deepak, M.; Tony, K.; Atma, P.; Sajid, A.; Faik, H. Effect of Geometric Configuration of the Impeller on the Performance of Liquivac Pump: Single Phase Flow (Water). *Fluids* **2022**, *7*, 45. [CrossRef]
18. Yu, T.; Shuai, Z.; Jian, J.; Wang, X.; Ren, K.; Dong, L.; Li, W.; Jiang, C. Numerical study on hydrodynamic characteristics of a centrifugal pump influenced by impeller-eccentric effect. *Eng. Fail. Anal.* **2022**, *138*, 106395. [CrossRef]
19. Elyamin, G.R.H.A.; Bassily, M.A.; Khalil, K.Y.; Gomaa, M.S. Effect of impeller blades number on the performance of a centrifugal pump. *Alex. Eng. J.* **2019**, *58*, 39–48. [CrossRef]
20. Zhaoheng, L.; Fangfang, Z.; Faye, J.; Ruofu, X.; Ran, T. Influence of the hydrofoil trailing-edge shape on the temporal-spatial features of vortex shedding. *Ocean. Eng.* **2022**, *246*, 110645.
21. Menter, F.R. Zonal two equation k.ω turbulence models for aerodynamic flows. In Proceedings of the 23rd Fluid Dynamics, Plasmadynamics, and Lasers Conference, Orlando, FL, USA, 6–9 July 1993; p. 2906.
22. Lu, Z.; Xiao, R.; Tao, R.; Li, P.; Liu, W. Influence of guide vane profile on the flow energy dissipation in a reversible pump-turbine at pump mode. *J. Energy Storage* **2022**, *49*, 104161. [CrossRef]
23. Fang, X.; Hou, Y.; Cai, Y.; Chen, L.; Lai, T.; Chen, S. Study on a high-speed oil-free pump with fluid hydrodynamic lubrication. *Adv. Mech. Eng.* **2020**, *12*, 1687814020945463. [CrossRef]
24. Stuparu, A.; Baya, A.; Bosioc, A.; Anton, L.; Mos, D. Experimental investigation of a pumping station from CET power plant Timisoara. *IOP Conf. Ser. Earth Environ. Sci.* **2019**, *240*, 032018. [CrossRef]
25. Hassan, S.M.; Salman, S. Effects of impeller geometry modification on performance of pump as turbine in the urban water distribution network. *Energy* **2022**, *255*, 124550.
26. Liu, Y.; Xia, Z.; Deng, H.; Zheng, S. Two-Stage Hybrid Model for Efficiency Prediction of Centrifugal Pump. *Sensors* **2022**, *22*, 4300. [CrossRef] [PubMed]
27. Francesco, P.; Maurizio, G. An Operative Framework for the Optimal Selection of Centrifugal Pumps As Turbines (PATs) in Water Distribution Networks (WDNs). *Water* **2022**, *14*, 1785.
28. Parkes, A.I.; Sobey, A.J.; Hudson, D.A. Physics-based shaft power prediction for large merchant ships using neural networks. *Ocean. Eng.* **2018**, *166*, 92–104. [CrossRef]
29. Wang, X.; Zhang, J.; Li, Z. Numerical Simulation of Internal Flow Field of Self-Designed Centrifugal Pump. In Proceedings of the 2021 International Conference on Fluid and Chemical Engineering (ICFCE 2021), Wuhan, China, 29–30 October 2021; pp. 158–166.
30. Sun, Z.; Yu, J.; Tang, F. The Influence of Bulb Position on Hydraulic Performance of Submersible Tubular Pump Device. *J. Mar. Sci. Eng.* **2021**, *9*, 831. [CrossRef]

Article

Simulation of Internal Flow Characteristics of an Axial Flow Pump with Variable Tip Clearance

Jiantao Shen [1], Fengyang Xu [2], Li Cheng [1,*], Weifeng Pan [3], Yi Ge [4], Jiaxu Li [1] and Jiali Zhang [4]

[1] College of Hydraulic Science and Engineering, Yangzhou University, Yangzhou 214000, China; shenjiantao888@163.com (J.S.); lijiaxu_yzu@163.com (J.L.)
[2] Jiangsu Zhenjiang Jianbi Pumping Station Management Office, Zhenjiang 212006, China; xufengyang06@163.com
[3] Luoyun Water Conservancy Project Management Divison in Jiangsu Province, Suqian 223800, China; jssqpwf@163.com
[4] Jurong Water Conservancy Bureau in Jiangsu Province, Jurong 212499, China; geyi1988@126.com (Y.G.); jrfxfh@126.com (J.Z.)
* Correspondence: chengli@yzu.edu.cn

Abstract: This study investigated the influence of the change in blade tip clearance on the internal flow characteristics of a vertical axial flow pump. Taking the actual running vertical axial flow pump of a pumping station as the research object, based on the SST k-ω turbulent flow model, the numerical simulation technology was used to study the effects of different tip clearances on the pressure, turbulent kinetic energy, Z–X section pressure and flow state of the impeller at the middle section. Furthermore, the impact of clearance layer tip leakage was also analyzed. Unsteady calculations of flow characteristics under the design conditions were performed. The research results showed that the variation trend of the pressure in the impeller was basically the same under different tip clearance values. With the increase in the clearance value, the pressure gradient along the water inlet direction of the blade decreased and the leakage vorticity increased. Observing the leakage vorticity distribution of the gap layer under the flow condition of $0.6Q_0$, it was found that when the tip clearance was smaller than 1 mm, the leakage flow was small and easily assimilated by the mainstream, and the leakage flow and mainstream had a certain ability to compete, which caused adverse effects on the performance of the pump device. The pressure pulsation characteristics showed that the leakage flow caused by the tip clearance caused a high-frequency distribution, and the clearance obviously influenced the pressure pulsation characteristics.

Keywords: tip clearance; vertical axial flow pump; whole channel numerical simulation; pressure pulsation; leakage vortex

Citation: Shen, J.; Xu, F.; Cheng, L.; Pan, W.; Ge, Y.; Li, J.; Zhang, J. Simulation of Internal Flow Characteristics of an Axial Flow Pump with Variable Tip Clearance. *Water* **2022**, *14*, 1652. https://doi.org/10.3390/w14101652

Academic Editors: Ran Tao, Changliang Ye and Xijie Song

Received: 19 April 2022
Accepted: 20 May 2022
Published: 22 May 2022

Publisher's Note: MDPI stays neutral with regard to jurisdictional claims in published maps and institutional affiliations.

Copyright: © 2022 by the authors. Licensee MDPI, Basel, Switzerland. This article is an open access article distributed under the terms and conditions of the Creative Commons Attribution (CC BY) license (https://creativecommons.org/licenses/by/4.0/).

1. Introduction

In the areas along the rivers in China, in order to meet the requirements of water transfer and irrigation, there are many vertical axial flow pumping stations [1,2]. These axial flow pumping stations have the characteristics of large flow and low lift [3], have been widely used in the eastern route of the South-to-North Water Diversion Project, and have achieved great social and economic benefits [4]. In the design of the vertical axial flow pump, due to the way water is pumped [5] and in order to prevent the impeller from scraping and colliding with the pump casing when the impeller rotates at high speed, it is necessary to leave a certain gap between the top of the impeller blade and the pump casing, which is the tip clearance. When the impeller is working, there is a pressure difference between the working surface and the back of the blade, and the existence of the pressure difference will lead to the generation of a leakage flow at the clearance position at the top of the blade. The interaction between the leakage flow and the main flow not only affects the pump performance but also induces abnormal vibration and noise, and even causes safety hazards and affects the normal operation of the pumping station [6].

Domestic and foreign experts and scholars studied the influence of tip clearance flow laws and the performance of impeller machinery. Zhang Desheng et al. [7] studied the pressure difference distribution, leakage, tip leakage vortex intensity and inlet axial velocity distribution in the tip clearance area of an inclined flow pump under different tip clearances. Shi Weidong et al. [8] used the SMPLEC algorithm to simulate different clearances under different working conditions and analyzed the axial velocity and circulation distribution of an impeller outlet in detail. Zhang et al. [9] revealed the gap flow field structure and leakage vortex evolution of a semi-open centrifugal pump by changing the gap according to the numerical calculation of the whole flow channel of the semi-open impeller centrifugal pump. Li Hui et al. [10] used the delayed detached vortex method to simulate a dynamic blade with a tip clearance and then studied the turbulent characteristics of the leakage area through the turbulent kinetic energy distribution and Lumley triangle. Then, the POD decomposition method was used to decompose the flow field in the leakage area, and finally, the loss analysis of the tip leakage flow was carried out. Lu Jinling et al. [11] carried out a full flow channel numerical simulation of a semi-open centrifugal pump using Fourier transform (FFT) to convert the time domain value of each monitoring point to a frequency domain value and analyzed the correlation mechanism between leakage vortex trajectory and blade load, as well as the spectral characteristics of the leakage vortex. Li et al. [12] studied the internal flow characteristics of the tip clearance of tubular turbines under off-design conditions and analyzed the pressure and velocity vector distribution of the internal flow field in the tip clearance, as well as the axial velocity and turbulent kinetic energy distribution characteristics in the tip clearance. Liu et al. [13] observed the development and trajectory of the leakage vortex by changing the tip clearance and performed a spectral analysis. When the tip clearance increased from 0 mm to 10 mm, the maximum amplitude of the pressure pulsation in the impeller increased sharply. Due to the increase in leakage flow, the main frequency increased from 145 Hz to 184 Hz. Kan et al. [14] used a standard k-epsilon turbulence model to simulate the flow characteristics of a self-developed helical axial flow pump. Pressure, streamline and turbulent kinetic energy analyses of the leakage flow (TLF) in a helical axial flow pump were undertaken, revealing the effect of tip clearance on the flow behavior and boosting performance of a helical axial flow pump. Shen et al. [15] used a computational fluid dynamics method to study the effect of different tip clearance widths on the tip flow dynamics and main flow characteristics of axial flow pumps. The distribution 0.7Q (BEP) of turbulent kinetic energy, average axial velocity and average vorticity at a specific flow rate was analyzed, and it was found that the flow structure of the tip vortex and its transmission strongly depended on the tip gap width. The efficiency and head of the pump both increase the energy loss as the tip clearance increases. Feng Jianjun et al. [16] studied the pressure pulsation of an axial flow pump under different tip clearances and proposed a new method to determine the pressure pulsation of the grid node in a pump based on pressure statistics. It was found that the existence of a tip clearance enlarged the pressure pulsation from the hub to the rim of the impeller. Li Yibin et al. [17] studied the pressure pulsation characteristics of different tip clearance regions of a diagonal flow pump under the condition of a small flow rate, and the influence on the transient operation stability of the diagonal flow pump was revealed. Through the analysis of the pressure pulsation spectrum, the internal relationship between the R_{TC} and the pressure pulsation in the vicinity of the tip clearance area was understood. The results show that selecting an appropriately small R_{TC} can improve the overall hydraulic performance of the oblique flow pump. Zhang Hua et al. [18] studied the influence degree and mechanism of the blade tip clearance on the internal and external characteristics of the centrifugal pump, a special adjustment mechanism for the blade tip clearance was designed, and the external characteristics test and a pressure pulsation test were carried out at the same time. The results show that the ratio of the tip clearance to the average diameter of the impeller should be between 0.13% and 0.22%. This provides a reference for the vibration and noise reduction of a semi-open screw centrifugal pump. Li Rennian et al. [19] studied the influence of the tip clearance on the inlet pressure pulsation characteristics of the diagonal flow pump, where the results show that with the increase of the tip clearance, the lift of the diagonal flow pump decreases gradually. The amplitude of pressure pulsation in the mainstream area of the impeller inlet is small, while the amplitude of pressure pulsation in the near-wall area is large. The larger tip clearance can reduce the pressure pulsation amplitude at the impeller inlet, which is beneficial for improving the operation stability of the model oblique flow pump. Li Yaojun et al. [20] studied the unsteady flow characteristics of the axial flow pump rim area under different tip clearances, where the results showed that when the tip clearance increased from $0.001D_2$ to $0.003D_2$, the pump head and efficiency decreased by 6.2% and 5.6%, respectively. When the clearance value is greater than $0.001D_2$, the primary leakage vortex in the rim clearance developed to the front face of the adjacent blade and a large number of secondary leakage vortices are generated in the clearance area. Guelich, J. and Ulanicki, B. et al. [21,22] studied the efficiency and power characteristics of a pump group, which were simulated by changing the Reynolds number and considering the constant or variable speed of the pump.

In this study, a full flow channel numerical calculations were carried out for the vertical axial flow pump [23], and the influence of the tip clearance change on the total pressure, turbulent kinetic energy distribution, Z–X section pressure, flow state, and the leakage flow and leakage volume of the gap layer were analyzed in detail. Numerical simulations of the unsteady flow in the axial flow pump with different tip clearance sizes were carried out to study the effect of the clearance size on the leakage vortex shape of the rim clearance and to analyze the pressure pulsation characteristics of the clearance layer to obtain the optimal blade rim. The variation law of the side and inlet and outlet pressure pulsations with different rim gap sizes theoretically revealed the gap flow field structure and leakage flow evolution of the vertical axial flow pump with a changing gap. This provides reference significance for more efficient operation and management of vertical axial flow pump stations [24,25].

2. Numeral Calculations

2.1. Vertical Axial Flow Pump Parameters and Research Plan

This study took a vertical bidirectional axial flow pump station as the prototype for the 3D modeling. The pump station adopted the 2500ZLQ-20-3.0 vertical axis open axial flow pump produced by a factory. The designed flow rate of the pump was $Q_0 = 20\ m^3/s$, the impeller diameter was 2.50 m, the number of impeller blades was 3, the number of guide blades was 7, the specific speed was 1250 and the impeller speed was 150 r/min.

In order to study the change in the internal flow characteristics of the pump under different clearance sizes and to ensure that the external characteristics of the prototype pump remained unchanged, the method of turning the impeller was adopted [26]. Four different tip clearances (d = 1 mm, 2.5 mm, 3.5 mm, 5 mm) were selected to simulate the flow under five working conditions. Through comparative analysis, the internal flow of vertical axial flow pump with different tip clearances was expounded. Figure 1a below is the flow chart of the research plan. Figure 1b is the tip clearance diagram of the vertical axial flow pump.

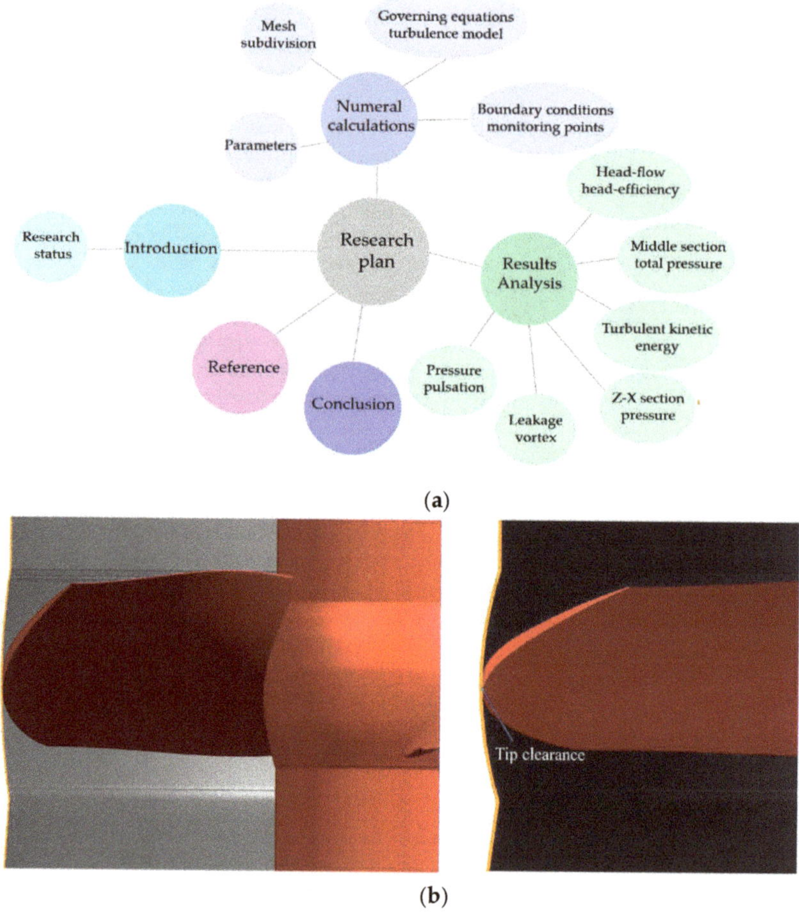

(**a**)

(**b**)

Figure 1. (**a**) Research program flow chart. (**b**) Three-dimensional schematic diagram of the tip clearance.

2.2. Mesh Subdivision

As shown in Figure 2a, the physical model of the computational domain in this study included the inlet section, the impeller section, the guide vane section and the outlet section. Considering the control of the overall number of mesh elements and the number of effective nodes, a hexahedral mesh was used for the divisions and the impeller section was refined to control the mesh quality.

Since the mesh size and quality have an important influence on the calculation results, by adjusting the mesh size of the impeller blade surface and the rim clearance area, the design flow conditions of the pump under different mesh schemes when the tip clearance was d = 2.5 mm were compared. The efficiency of the pump was compared with the test results, and a mesh-independent analysis was conducted. As shown in Table 1, the pump head and efficiency calculated using grid schemes A, B and C were basically the same. Considering the economy of numerical calculation, grid scheme B was adopted in this study. With different tip clearances, the total number of computational domain grids was about 1.3×10^7; the global mesh and impeller vane mesh are shown in Figure 2b.

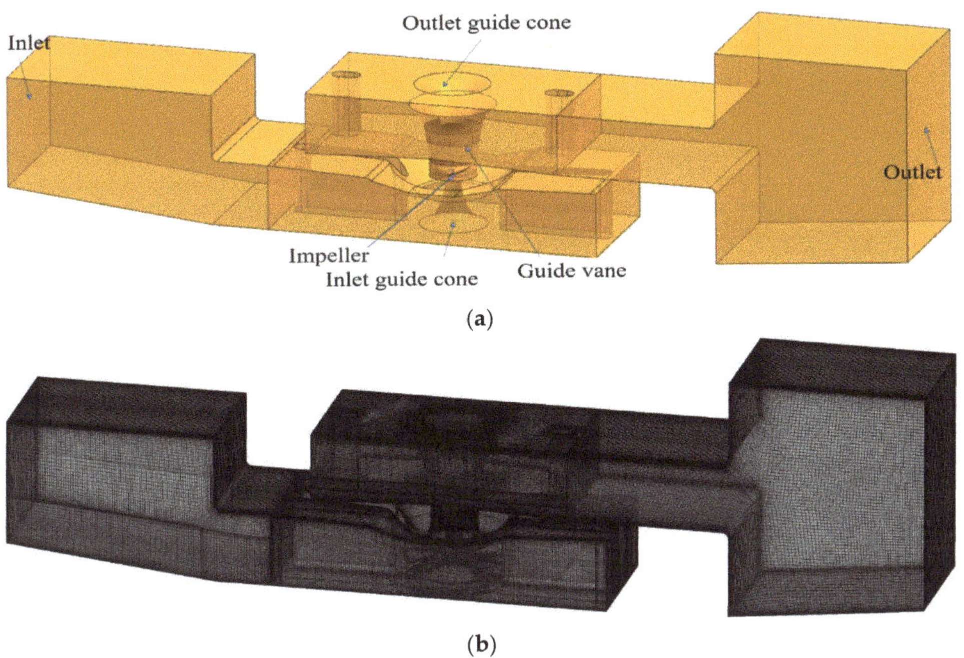

Figure 2. (**a**) Computational fluid domain of the whole channel. (**b**) Global grid.

Table 1. Grid independence.

Plan	Total Number of Grids	Efficiency Test Value /%	Efficiency Calculated Value /%	Efficiency Relative Error /%	Head Test Value /m	Head Calculated Value/m	Head Relative Error /m
A	10,697,069	61.61	64.25	4.29	1.90	1.9283	2.79
B	13,422,014	61.61	63.72	3.42	1.90	1.9003	1.31
C	15,342,176	61.61	63.34	2.81	1.90	1.8916	0.83

2.3. Governing Equations and Turbulence Model Selection

The internal flow of a pump is unsteady viscous flow, which can be described using Navier–Stokes equations [27]. The Reynolds-averaged Navier–Stokes (RANS) method was used to decompose the various characteristic variables of the turbulent flow into time-averaged Reynolds equations and fluctuating values, and then the turbulent viscosity coefficient was introduced to establish the turbulent model. For a general incompressible Newtonian fluid, the following control equations are used.

The mass conservation equation, also called the continuity equation, is as follows:

$$\frac{\partial \rho_f}{\partial t} + \nabla \cdot \left(\rho_f v\right) = 0$$

Momentum equation:

$$\frac{\partial \rho_f v}{\partial t} + \nabla \cdot \left(\rho_f v v - \tau_f\right) = f_f$$

where t is time, f_f is the volume force vector, ρ_f is the fluid density, v is the fluid velocity vector and τ_f is the shear force tensor given by

$$\tau_f = (-p + \mu \nabla \cdot v) I + 2\mu e$$

where p is the static pressure, μ is the dynamic viscosity and e is the velocity stress tensor:

$$e = \left(\nabla v + \nabla v^T\right)$$

The numerical simulation and experimental results showed that under the optimal conditions, the predicted external characteristic curve of the SST k-ω turbulence model was in good agreement with the experimental curve, where the head error was 4.688% [28]. Therefore, the SST k–ω was selected as the turbulence model to conduct the numerical simulation of the whole flow channel in the computational domain of the axial flow pump, where k is the turbulent kinetic energy, and the transport equation is:

$$\rho\frac{\partial(k)}{\partial t} + \rho\frac{\partial}{\partial x_j}(U_j k) = \frac{\partial}{\partial x_j}\left[\left(\mu + \frac{\mu_t}{\sigma_k}\right)\frac{\partial k}{\partial x_j}\right] + P_k - \beta'\rho k\omega$$

ω is the turbulent dissipation rate, and the related equation is

$$\frac{\partial(\rho\omega)}{\partial t} + \frac{\partial}{\partial x_j}(\rho U_j\omega) = \frac{\partial}{\partial x_j}\left[\left(\mu + \frac{\mu_t}{\sigma_\omega}\right)\frac{\partial \omega}{\partial x_j}\right] +$$
$$\alpha\frac{\omega}{k}P_k - \beta\rho\omega^2 + 2(1 - F_1)\rho\frac{1}{\sigma_{\omega 2}\omega}\frac{\partial k}{\partial x_j}\frac{\partial \omega}{\partial x_j}$$

In this formula, U_j is the vector velocity (m/s), P_k is the turbulent generation rate and μ_t is the turbulent viscosity (m^2/s).

2.4. Boundary Conditions and Arrangement of the Pressure Fluctuation Monitoring Points

The boundary conditions in this study were the velocity inlet and free outlet. The flow at the pump outlet was fully developed. The impeller speed was n = 150 r/min and the rotor–stator dynamic–static interface was a frozen rotor. The wall of each flow component adopted a smooth non-slip wall, The impeller surface and hub surface were set as moving walls without heat transfer, and the convergence accuracy was 10^{-4}. In the unsteady calculations, the time step was 1/40 of the impeller rotation period of 0.4 s, and the sampling time was 8 impeller rotation periods. Detailed parameters are shown in Table 2.

Table 2. Parameter settings for the constant and unsteady calculations.

Calculated Parameters	Settings	Calculated Parameters	Settings
Flow assumption	Incompressible	Static–static interface	GGI
Simulation type	Steady	Dynamic–static interface	Frozen rotor
Inlet boundary condition	Quality inlet	Wall condition	No slippage
Outlet boundary condition	Pressure outlet	Wall function	Scalable wall function
Impeller speed	150 r/min	Convergence accuracy	10^{-4}
Flow assumption	Incompressible	Impeller speed	150 r/min
Simulation type	Transient	Static–static interface	GGI
Time Step	0.01 s	Dynamic–static interface	Transient rotor stator
Total time	0.32 s	Wall conditions	No slippage
Inlet boundary condition	Quality inlet	Wall function	Scalable wall function
Outlet boundary Condition	Pressure outlet	Convergence accuracy	10^{-4}

In order to analyze the pressure fluctuation characteristics of a blade near the flange side, as well as the inlet and outlet, under the influence of different flange clearance leakage flow, the pressure fluctuation monitoring points were arranged as shown in Figure 3.

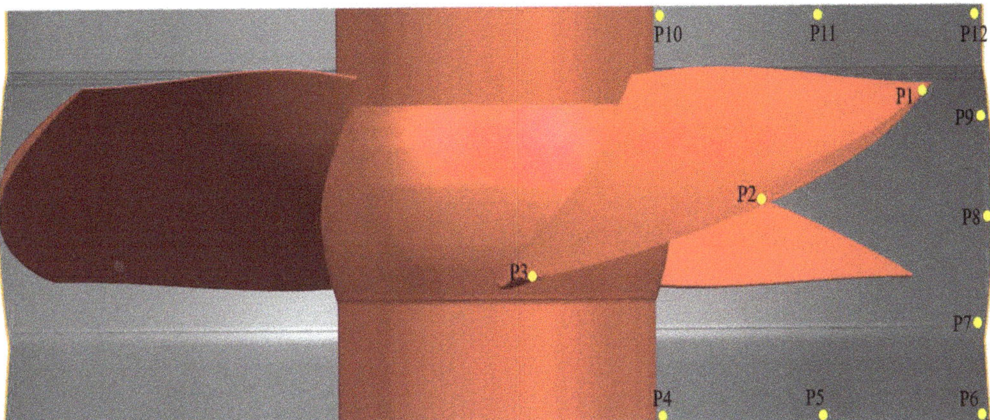

Figure 3. Schematic diagram of the monitoring point layout.

3. Calculation Results and Analysis

3.1. Head–Flow Curve and Head–Efficiency Curve

Figures 4 and 5 give the head–flow curve and the efficiency–flow curve under different clearances, respectively, where the YJ values represent the sizes of the different tip clearances in units of mm.

The test situation represented the head when the gap was 1 mm and evaluated the relationship between the efficiency and flow rate. It can be seen from Figure 4 that the test head value was generally higher than the numerical simulation head value, and decreased with the increase in the flow rate. However, when the flow rate increased, the drop rate was larger than the numerical model head, and there was a cross-over with the numerical model head value. The gap between the test and simulation was only about 0.2 m. Figure 5 shows that the efficiency of the test was lower than the simulation efficiency under the condition of a small flow rate, but the efficiency growth rate was larger than the simulation value. At $0.6Q_0$, the test efficiency gradually intersected with the simulation efficiency and began to exceed it. At $0.9Q_0$, the efficiency gap between the test and the simulation was the largest, reaching less than 2%. It shows that the efficiency gap was within the controllable range, and the numerical simulation was basically consistent with the experimental situation.

This shows that the digital–analog calculation was closer to the actual situation. It can be seen from Figure 5 that there was no linear correlation between the drop in the lift and the increase in the clearance. In the process of increasing the clearance from 1 mm to 5 mm, with the increase in the clearance, the change in the lift and efficiency was relatively small. When the gap was small, the leakage was small; the leakage was very small relative to the mainstream, and the impact on the mainstream was negligible, making the efficiency higher. It can be seen from Figure 5 that with the gradual increase in the gap, the leakage had a certain influence on the mainstream and the influence on the flow field increased, which caused the head and efficiency drops to be significantly larger. When the clearance continued to increase (referring to the 3.5 mm–5 mm stage), the leakage volume also continued to increase, but the increase in the tip clearance was moderate compared with the previous one. The difference in leakage flow was limited; therefore, the drop in the head and efficiency was also smaller than before.

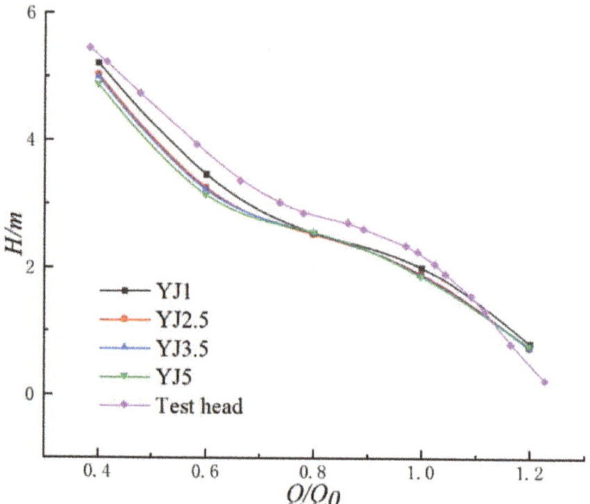

Figure 4. Head–flow curve.

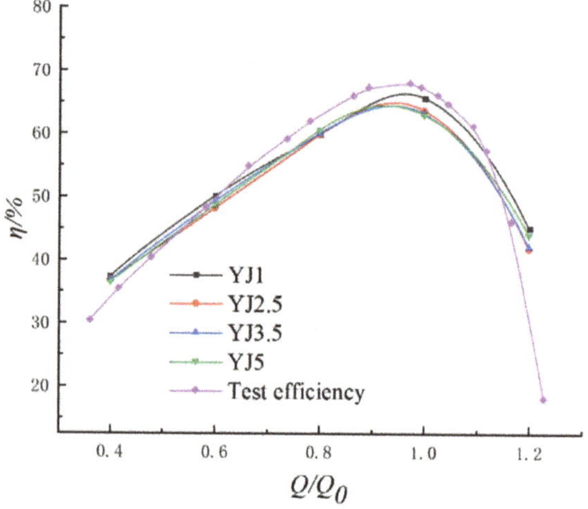

Figure 5. Efficiency–flow curve.

3.2. *Effect of Clearance Variation on Total Pressure Distribution of Impeller Middle Section*

Figure 6 shows the total pressure distribution of the middle section of the impeller under different tip clearances. It can be seen from Figure 6 that the pressure changed in the same trend under different tip clearances. The diffusion and the pressure at the outlet were small, and the degree of pressure diffusion was relatively uniform in this section, but at the edge of the front end of the impeller, affected by the gap leakage vortex, a local high-pressure area appeared. There were high-pressure areas on both the working face and the back of the blade, which were mainly due to the existence of the tip clearance. The high-pressure fluid on the working face of the blade passed through the tip clearance and directly acted on the back of the blade. With the increase in the clearance value, the pressure gradient along the water inlet direction of the blade decreased. As the leakage vorticity increased, the work done by the water on the blade decreased. When the tip clearance increased from 1 mm to 5 mm and the tip clearance was small, the leakage vorticity was small, and it was easier to generate a local high pressure at the edge of the inlet direction.

When the tip clearance value was large, the high-pressure fluid crossed the gap. It acted on the back of the blade, which reduced the workability of the water flow for the blade.

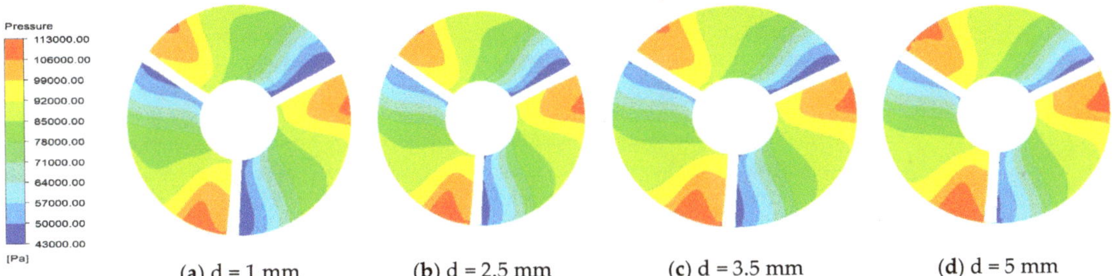

Figure 6. Total pressure distribution of the middle section of the impeller under the design conditions.

3.3. Effect of the Clearance Variation on the Turbulent Kinetic Energy Distribution in the Middle Section of Impeller

Figure 7 shows the distribution of the turbulent kinetic energy in the middle section of the impeller under the design conditions. It can be seen from the figure that the area with a large turbulent kinetic energy was concentrated at the edge of the impeller, and the turbulent kinetic energy was particularly large near the tip of the blade. This was because the high-pressure fluid leaked from the gap and crossed the working face to the back to form a mixing loss zone. This affected the normal flow of the fluid in the flow channel, where a region of high turbulent kinetic energy was created. In the four different tip clearance cases, the tip leakage flow increased with the increase in the tip clearance size, and the extreme value of the leakage flow increased when the turbulent kinetic energy region increased. Combined with the analysis of the total pressure distribution in the middle section of the impeller in Figure 6, it can be seen that the pressure gradient in the flow channel along the water flow direction was reduced and the influence of the leakage flow changed, resulting in the formation of unstable turbulent pulsations inside the flow channel. The interaction of fluid molecules, friction and collisions aggravated the internal energy loss of the fluid such that a part of the internal energy was converted into heat energy and conducted out, thereby reducing the efficiency of the pump device.

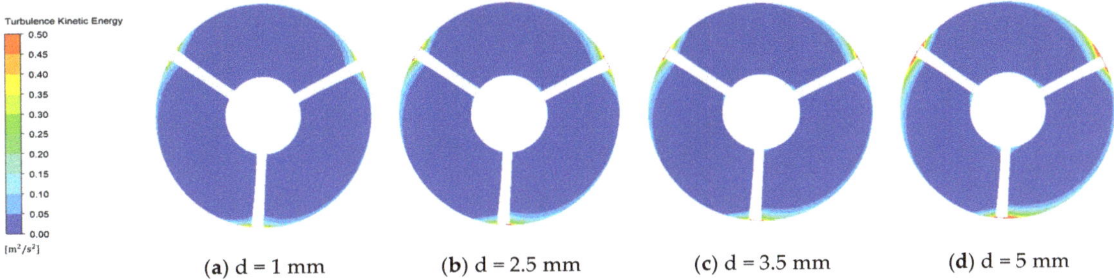

Figure 7. Distribution of the turbulent kinetic energy in the middle section of the impeller under the design conditions.

3.4. Effect of the Gap Change on the Pressure and Flow State of the Z–X Section

Figure 8 shows the pressure distribution of Z–X section with different tip clearances under the design conditions. It can be seen that under different tip clearance values, high-pressure fluid entered from the bottom inlet, passed through the impeller guide vane in turn and flowed out from the upper outlet. The total pressure of the Z–X section increased gradually along the inlet direction and reached the maximum at the outlet, and there was a local high-pressure area at the outlet section of the impeller. The total pressure fluctuation at the inlet position of the impeller was not obvious. When the high-pressure fluid reached the blade working face, the isobaric line began to change significantly and was accompanied by large fluctuations. The pressure fluctuations were mostly concentrated in the impeller channel, which intensified the instability of the fluid flow. It can also be seen from Figure 8 that the pressure of the blade's working face was significantly greater than that of the back surface, and even the pressure of some regions was close to zero at the back surface of the blade. This was because the high-pressure fluid near the tip clearance of the impeller blade working face side leaked to the back of the blade through the clearance, and the working efficiency of the blade was reduced, which showed the trend of pressure reduction on the back of the blade.

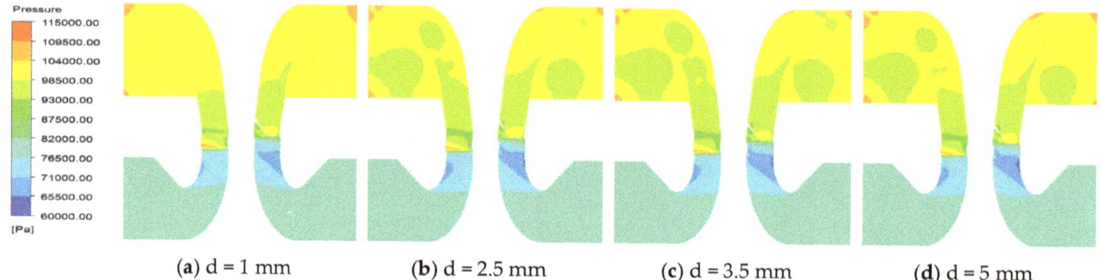

(**a**) d = 1 mm (**b**) d = 2.5 mm (**c**) d = 3.5 mm (**d**) d = 5 mm

Figure 8. Pressure distribution of the Z–X section under the design conditions.

As shown in Figure 9, due to the existence of the tip clearance, there was a significant backflow at the top of the impeller blade, flow around the outlet, flow pattern disorder and so on. The flow entered the impeller, and with the increase in tip clearance, the backflow in the impeller channel increased significantly. When the clearance was small, the leakage was small, resulting in a large pressure difference on both sides of the blade, which led to the formation of a flow vortex in the impeller passage. With the gradual increase in the tip clearance, the leakage increased, causing part of the fluid gathered in the working face to cross the clearance to form backflow. The larger the clearance was, the more backflow there was. By observing the above figure, it can be seen that the backflow caused by a certain gap (Figure 9b) could make the fluid flow smoother, which was sufficient to demonstrate that a certain gap (Figure 9b) could improve the flow pattern of the impeller channel.

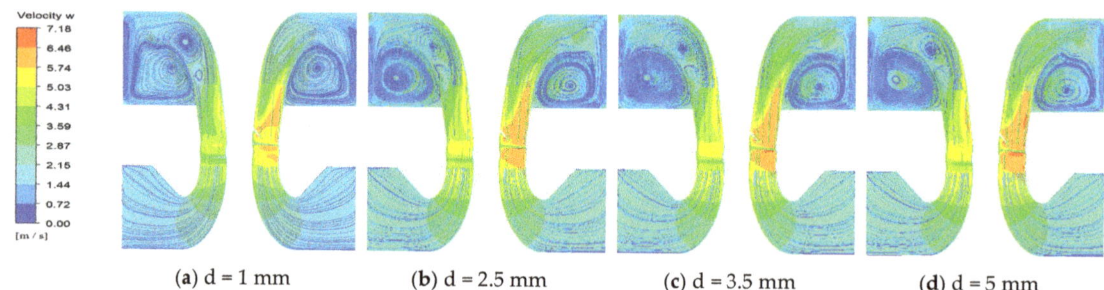

(**a**) d = 1 mm (**b**) d = 2.5 mm (**c**) d = 3.5 mm (**d**) d = 5 mm

Figure 9. Velocity distribution of the Z-X section under the design conditions.

3.5. Influence of Clearance Change on Tip Leakage of Clearance Layer

Figures 10 and 11 show the relative velocity vectors and leaky vortex 3D shapes under different tip clearances at a $0.6Q_0$ flow rate. When the tip clearance was 1 mm, the leakage of the tip head relative to other positions was large and the pressure difference between the working face and the back of the blade was also the largest. Since the tip clearance was very small relative to the impeller diameter and the collision between the leakage flow and the pump shell caused loss, the actual leakage flow was a small part relative to the mainstream and had little effect on the flow. Under the influence of the mainstream after the formation of the leakage flow, the leakage flow was mixed into the mainstream. The pump device had good performance under this small tip clearance.

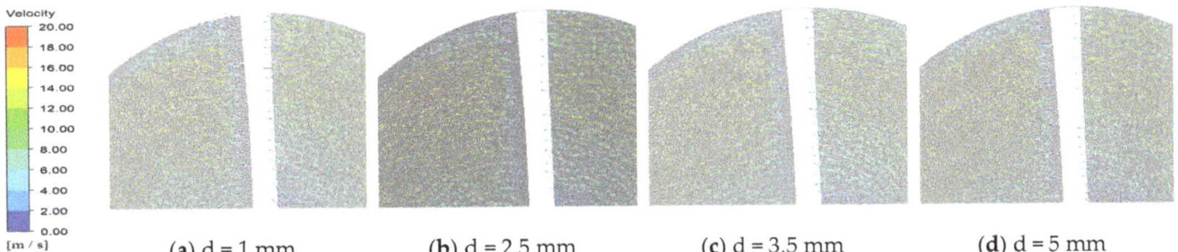

(**a**) d = 1 mm (**b**) d = 2.5 mm (**c**) d = 3.5 mm (**d**) d = 5 mm

Figure 10. Relative velocity vector diagram under the $0.6Q_0$ flow condition.

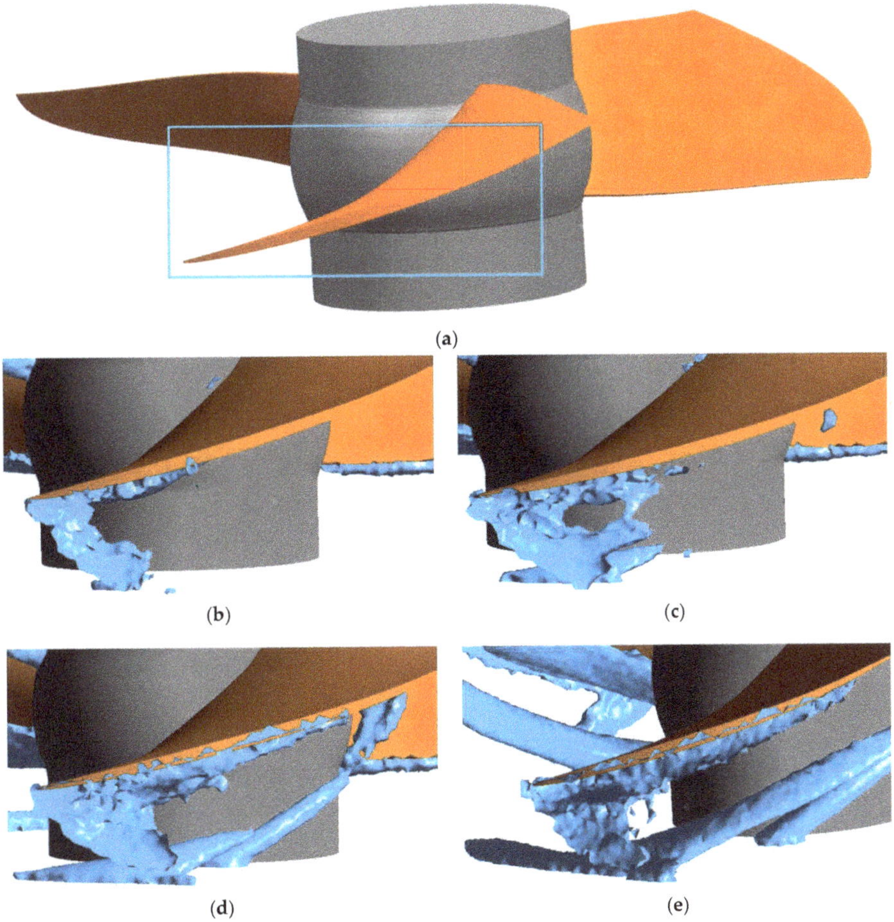

(**a**)

(**b**) (**c**)

(**d**) (**e**)

Figure 11. Leaky vortex 3D shape. (**a**) Observation angle; (**b**) d = 1 mm; (**c**) d = 2.5 mm; (**d**) d = 3.5 mm; (**e**) d = 5 mm.

As the tip clearance gradually increased, the leakage also gradually increased. Compared with Figure 11a, a relatively obvious leakage flow was formed in Figure 11b. At this time, the influence of the leakage flow on the mainstream also increased, but the leakage flow at this time was still relatively small and was still mixed into the mainstream, and thus, the leakage vortex disappeared. As the tip clearance was further expanded to 5 mm, the leakage volume was further increased and the leakage vortex intensified. The leakage flow and the mainstream had a certain ability to resist and interact. In the encounter position, the entrainment phenomenon occurred, resulting in the intensification of the vortex in the flow channel. At this time, the larger tip clearance had a negative impact on the performance of the pump device and caused concomitant damage to the blade.

3.6. Characteristics of the Pressure Pulsation in the Tip Clearance Region

In order to further reveal the influence of the tip clearance on the flow characteristics in the rim area of the vertical axial flow pump under the design conditions, the time domain and frequency domain characteristics were analyzed by monitoring the pressure pulsation characteristics in the tip clearance area. The amplitude of the pressure pulsation was described by the relative value of the double-amplitude peak-to-peak value of the mixing frequency in the time domain, and the spectral characteristics of the pressure pulsation were obtained by using the fast Fourier transform (FFT). In order to improve the frequency resolution of the FFT analysis, the sampling time was continuously calculated for four cycles, starting from the fourth impeller rotation cycle, and the static pressure data of the monitoring points in these four cycles were collected to analyze the flow field spectrum. The amplitude of the pressure pulsation was represented by the pressure pulsation coefficient Cp, which is expressed as follows:

$$C_p = \frac{\Delta P}{0.5\rho U^2}$$

where

$$U = \frac{\pi n D}{60}$$

In these formulae, ΔP is the difference between the instantaneous pressure and the average static pressure at the monitoring point (Pa); ρ is the fluid density (kg/m^3); D is the impeller diameter (m); U is the peripheral speed (m/s) and n is the pump speed (r/min).

Table 3 is the amplitude table of the monitoring points at the impeller under different tip clearances. From the data shown in this table, it can be seen that the main frequency of the monitoring points at the impeller was a multiple of the rotational frequency of the impeller. Because of the influence of the periodic rotation of the three blades of the impeller and the mutual interference of the static flow field at the inlet, the main frequency of most of the monitoring points at the impeller was the same as the blade frequency. In particular, the main frequency of the measuring points at the impeller rim and the inlet hub was the blade frequency. The outlet of the impeller was connected to the inlet of the guide vane; therefore, in addition to being affected by the main frequency of the impeller, some frequencies were also affected by the guide vane. Under different tip clearances, with the increase in the clearance, the amplitude first increased and then decreased, while the change trend of the position of the inlet and outlet and the axial monitoring position was basically the same; therefore, the three points at the rim were further analyzed.

Figure 12 shows the pressure pulsation characteristics of the monitoring point P3 near the impeller inlet for different tip clearances. It can be seen that the pressure pulsation at the monitoring point of the pressure pulsation at the impeller inlet had a relatively obvious law, which was similar to a sinusoidal waveform. There was a complete peak and trough in one cycle. When the tip clearance was small, the pressure variation range was large because the impact of the leakage flow was large under the same working conditions. When the tip clearance was small, there were many complex high-frequency components in the frequency domain. Since P3 was located at the leading edge of the tip, the pressure here was affected by the direct work done by the blade and also by the leakage flow, which caused the flow here to be more stable. It was complicated and exacerbated the instability of the flow.

Table 3. Amplitudes of the monitoring points at the impeller when the clearances were 1, 2.5, 3.5 and mm.

Location	Point Number	Main Frequency (Hz)	Amplitude
Three points of impeller rim clearance	P1 P2 P3	2.5 2.5 5.0	0.00345 0.00683 0.00817
Circumferential three points of impeller inlet	P4 P5 P6	7.5 2.5 2.5	0.00345 0.00393 0.00631
Three axial points at the impeller rim	P7 P8 P9	2.5 2.5 17.5	0.00540 0.00271 0.00340
Circumferential three points of impeller outlet	P10 P11 P12	7.5 17.5 17.5	0.00715 0.00928 0.01403
Three points of impeller rim clearance	P1	2.5	0.00322
	P2	2.5	0.00689
	P3	2.5	0.01339
Circumferential three points of impeller inlet	P4	7.5	0.00259
	P5	2.5	0.00390
	P6	2.5	0.00655
Three axial points at the impeller rim	P7	2.5	0.00556
	P8	2.5	0.00223
	P9	17.5	0.00381
Circumferential three points of impeller outlet	P10	7.5	0.00800
	P11	7.5	0.01643
	P12	17.5	0.01342
Three points of impeller rim clearance	P1	2.5	0.00374
	P2	2.5	0.00733
	P3	2.5	0.01285
Circumferential three points of impeller inlet	P4	7.5	0.00285
	P5	2.5	0.00406
	P6	2.5	0.00664
Three axial points at the impeller rim	P7	2.5	0.00571
	P8	2.5	0.00233
	P9	17.5	0.00376
Circumferential three points of impeller outlet	P10	7.5	0.00834
	P11	7.5	0.01673
	P12	17.5	0.01316
Three points of impeller rim clearance	P1	2.5	0.00395
	P2	2.5	0.00751
	P3	2.5	0.01189

Table 3. *Cont.*

Location	Point Number	Main Frequency (Hz)	Amplitude
Circumferential three points of impeller inlet	P4	7.5	0.00290
	P5	2.5	0.00402
	P6	2.5	0.00627
Three axial points at the impeller rim	P7	2.5	0.00553
	P8	2.5	0.00234
	P9	17.5	0.00364
Circumferential three points of impeller outlet	P10	7.5	0.00848
	P11	7.5	0.01679
	P12	17.5	0.01257

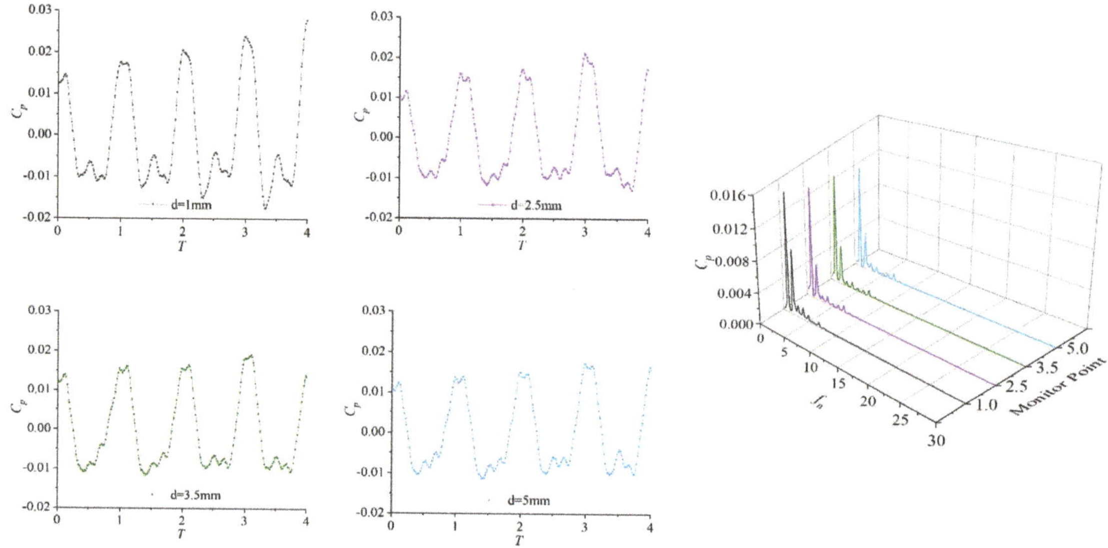

(**a**) Time domain diagram (**b**) Frequency domain plot

Figure 12. Pressure pulsation characteristics of the tip inlet edge point P3.

Figure 13 shows the pressure pulsation characteristics of the monitoring point P2 in the middle of the blade tip. It can be seen that the four different blade tip clearances showed similar regular changes, but the amplitudes were slightly different; its frequency domain characteristics also had complex components, but they were not as chaotic as the inlet end, and the middle of the blade tip was affected by the periodic rotation of the blade and leakage flow.

Figure 14 shows the pressure pulsation characteristics of the monitoring point P1 at the tip outlet. It can be seen that the regularity was not as obvious as before in the case of different tip clearances since there were no obvious periodic peaks and valleys, showing a certain amount of randomness; its frequency domain also had a large number of high-frequency parts, but it was not as good as the inlet and the middle of the tip, and as the tip clearance increased, the high-frequency part decreased, which was caused by interference from the leakage flow.

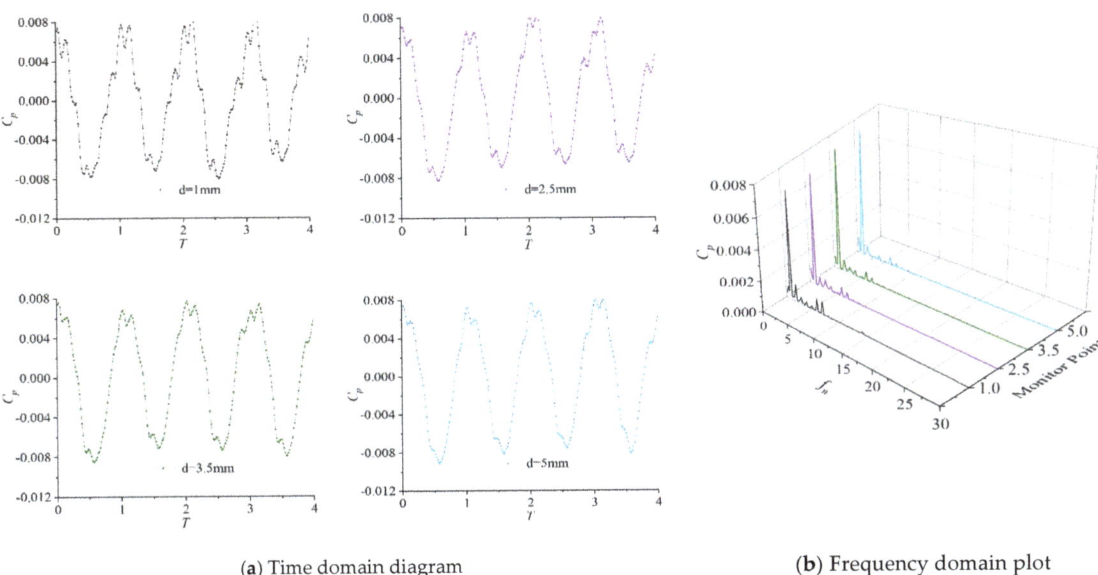

(**a**) Time domain diagram (**b**) Frequency domain plot

Figure 13. Pressure pulsation characteristics at point P2 in the middle of the blade tip.

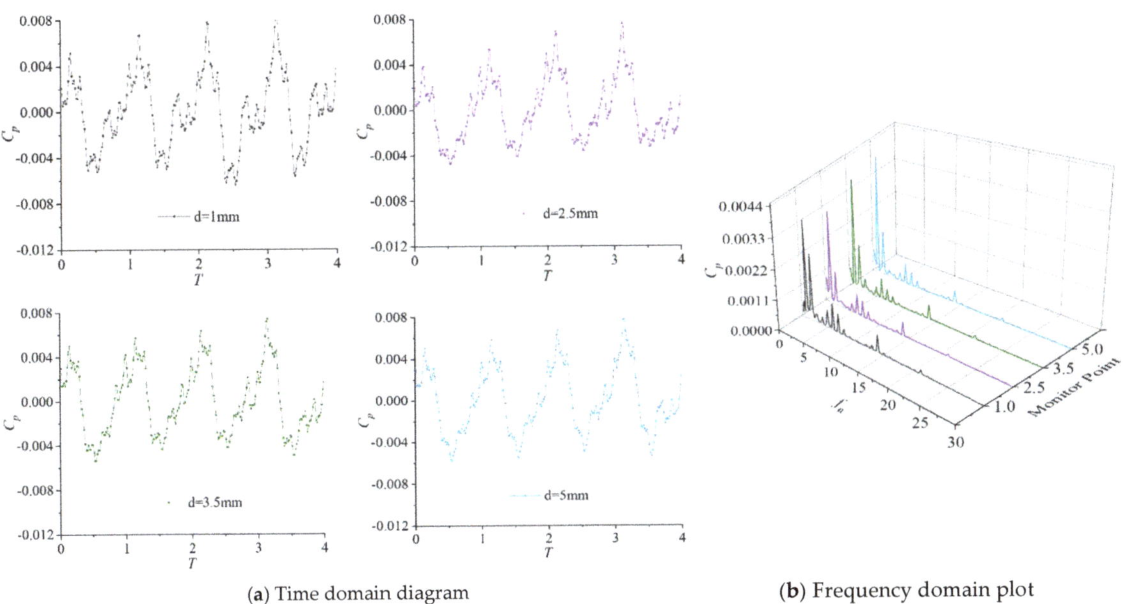

(**a**) Time domain diagram (**b**) Frequency domain plot

Figure 14. Pressure pulsation characteristics of the tip outlet edge point P1.

4. Conclusions

(1) As the tip clearance increased, the lift and efficiency decreased, but there was no linear relationship between the degree of the decrease and the change in the clearance. Furthermore, in the process of increasing the flow rate, the difference in the head under the size of the tip clearance first decreased and then increased, but the difference in the

head under the condition of a small flow was larger than that under the condition of a large flow. In the relationship between the efficiency and flow rate, from a small flow condition to a large flow condition, the efficiency first increased and then decreased, forming a saddle region, and the pump performance dropped sharply after the flow rate exceeded the design condition.

(2) In the Z–X section, there was an obvious backflow at the top of the impeller blade, and there was a flow around the outlet and flow pattern disorder. When the gap value was 2.5 mm, a certain amount of reflux caused the fluid flow to be smoother, indicating that a certain gap could improve the flow pattern of the channel.

(3) Under different tip clearance values, the variation trend of the pressure in the impeller was basically the same. However, the pressure fluctuation changed obviously at the flange clearance, and the pressure fluctuation amplitude was from 0.00322 to 0.01339.

(4) The pressure pulsation characteristics showed that the axial flow pump showed strong unsteady characteristics under the design conditions. The leakage flow caused by the tip clearance aggravated the flow instability and had a significant impact on the pressure pulsation characteristics in the clearance area.

Author Contributions: Data curation, L.C.; Formal analysis, J.S.; Methodology, F.X. and W.P.; Writing—original draft, J.S. and J.L.; Writing—review and editing, L.C., J.S., Y.G. and J.Z.; Supervision, L.C. All authors have read and agreed to the published version of the manuscript.

Funding: This research was funded by the National Natural Science Foundation of China (grant no. 51779214), a project funded by the Priority Academic Program Development of Jiangsu Higher Education Institutions (PAPD), the Key Project of Water Conservancy in Jiangsu Province (grant no. 2020030 and 2020027) and the Jiangsu Province South-to-North Water Transfer Technology Research and Development Project (SSY-JS-2020-F-45).

Institutional Review Board Statement: Not applicable.

Informed Consent Statement: Not applicable.

Data Availability Statement: Data on the analysis and reporting results during the study can be obtained by contacting the authors.

Acknowledgments: The authors thank the College of Hydraulic Science and Engineering, Yangzhou University. The author is very grateful for the discussions with Li Cheng and Jiao Weixuan. A huge amount of thanks is due to the editor and reviewers for their valuable comments to improve the quality of this paper.

Conflicts of Interest: The authors declare no conflict of interest.

References

1. Lu, X.; Tong, H.; Feng, J. Improvement of the influent flow state of the intake pool of the urban water pumping station. *Drain. Irrig. Mach.* **2007**, *25*, 24–28. (In Chinese) [CrossRef]
2. Hu, X.; Liu, J.; Wu, J.; Ma, Q. Analysis of the impact of drainage from the pumping stations along the middle and lower reaches of the Yangtze River on the flood control of the mainstream. *People's Yangtze River* **2020**, *51*, 172–178. (In Chinese) [CrossRef]
3. Liu, C. Analysis of technological innovation and development of axial flow pump system. *J. Agric. Mach.* **2015**, *46*, 49–59. (In Chinese) [CrossRef]
4. Zhao, C. Overview of the South-to-North Water Diversion Project. *Water Conserv. Constr. Manag.* **2021**, *41*, 5–9. (In Chinese) [CrossRef]
5. Mondal, P.; Mukherjee, S. A New Analytical Approach to Predict the Operating Point of a Pumping System Having Groups of Different Types of Radial-Flow Pumps in Parallel and the Resulting Flow Division in the Piping Network. *J. Inst. Eng.* **2012**, *93*, 83–91. [CrossRef]
6. Xie, K. Analysis of the status quo of real-time operation status assessment technology of pumping stations. *Think Tank Times* **2018**, *48*, 199+204. (In Chinese)
7. Zhang, D.; Shen, X.; Dong, Y.; Wang, C.; Liu, A.; Shi, W. Numerical simulation of internal flow characteristics of oblique flow pumps with different tip clearances. *Chin. J. Drain. Irrig. Mech. Eng.* **2020**, *38*, 757–763. (In Chinese) [CrossRef]
8. Shi, W.; Zhang, H.; Chen, B.; Zhang, D.; Zhang, L. Numerical calculation of the internal flow field of an axial flow pump with different tip clearances. *Chin. J. Drain. Irrig. Mech. Eng.* **2010**, *28*, 374–377+406. (In Chinese)

9. Zhang, Q.; Yang, J.; Li, H.; Yan, S.; Li, Y. Research on flow characteristics of variable tip clearance in semi-open impeller centrifugal pump. *Hydraul. Pneum. Seal.* **2021**, *41*, 6. (In Chinese)
10. Li, H.; Su, X.; Yuan, X. Research on turbulent characteristics of tip clearance flow based on DDES simulation. *J. Eng. Thermophys.* **2021**, *42*, 342–348. (In Chinese)
11. Lu, J.; Guo, L.; Wang, L.; Wang, W.; Guo, P.; Luo, X. Research on unsteady flow characteristics of tip clearance of half-open impeller centrifugal pump. *J. Agric. Mach.* **2019**, *50*, 163–172. (In Chinese) [CrossRef]
12. Li, Z.; He, F.; Pan, S. Internal flow analysis of tubular turbine blade tip clearance under partial working conditions. *Therm. Power Eng.* **2021**, *36*, 16–23. (In Chinese) [CrossRef]
13. Liu, Y.; Tan, L.; Hao, Y. Energy performance and flow patterns of a mixed-flow pump with different tip clearance sizes. *Energies* **2017**, *10*, 191. [CrossRef]
14. Kan, N.; Liu, Z.; Shi, G. Effect of Tip Clearance on Helico-Axial Flow Pump Performance at Off-Design Case. *Processes* **2021**, *9*, 1653. [CrossRef]
15. Shen, S.; Qian, Z.; Ji, B. Numerical investigation of tip flow dynamics and main flow characteristics with varying tip clearance widths for an axial-flow pump. *Proc. Inst. Mech. Eng. Part A J. Power Energy* **2019**, *233*, 476–488. [CrossRef]
16. Feng, J.; Luo, X.; Guo, P. Influence of tip clearance on pressure fluctuations in an axial flow pump. *J. Mech. Sci. Technol.* **2016**, *30*, 1603–1610. [CrossRef]
17. Li, Y.; Bi, Z.; Li, R.; Hu, P. Numerical analysis of pressure pulsation characteristics of oblique flow pump with different tip clearances. *J. Hydraul. Eng.* **2015**, *46*, 497–504. (In Chinese) [CrossRef]
18. Zhang, H.; Chen, B.; Wang, B.; Shi, C.; Shen, D. Effect of tip clearance on internal pressure pulsation of screw centrifugal pump. *Chin. J. Agric. Eng.* **2017**, *33*, 84–89. (In Chinese) [CrossRef]
19. Li, R.; Hu, P.; Li, Y.; Bi, Z.; Zhou, D. Numerical analysis of the influence of blade tip clearance on inlet pressure pulsation of oblique flow pump. *Chin. J. Drain. Irrig. Mech. Eng.* **2015**, *33*, 560–565. (In Chinese) [CrossRef]
20. Li, Y.; Shen, J.; Hong, Y.; Liu, Z. Numerical prediction of the influence of blade tip clearance on pressure pulsation of axial flow pump rim. *J. Agric. Mach.* **2014**, *45*, 59–64+58. (In Chinese) [CrossRef]
21. Guelich, J. Effect or Reynolds Number and Surface Roughness on the Efficiency of Centrifugal Pumps. *J. Fluids Eng.* **2003**, *125*, 670–679. [CrossRef]
22. Ulanicki, B.; Kahler, J.; Coulbeck, B. Modeling the Efficiency and Power Characteristics of a Pump Group. *J. Water Resour. Plan. Manag.* **2008**, *134*, 88–93. [CrossRef]
23. Shi, L.; Zhang, W.; Jiao, H.; Tang, F.; Wang, L.; Sun, D.; Shi, W. Numerical simulation and experimental study on the comparison of the hydraulic characteristics of an axial-flow pump and a full tubular pump. *Renew. Energy* **2020**, *153*, 1455–1464. [CrossRef]
24. Nm, A.; Gk, A.; Dh, A. A real-time energy management strategy for pumped hydro storage systems in farmhouses—ScienceDirect. *J. Energy Storage* **2020**, *32*, 101928. [CrossRef]
25. Luna, T.; Ribau, J.; Figueiredo, D. Improving energy efficiency in water supply systems with pump scheduling optimization. *J. Clean. Prod.* **2019**, *213*, 342–356. [CrossRef]
26. Shi, L.; Zhu, J.; Tang, F.; Wang, C. Multi-Disciplinary Optimization Design of Axial-Flow Pump Impellers Based on the Approximation Mode. *Energies* **2020**, *13*, 779. [CrossRef]
27. Yang, Y.; Zhou, L.; Bai, L.; Xu, H.; Lv, W.; Shi, W.; Wang, H. Numerical Investigation of Tip Clearance Effects on the Performance and Flow Pattern Within a Sewage Pump. *J. Fluids Eng.* **2022**, *144*, 081202. [CrossRef]
28. Zhang, D.; Wu, S.; Shi, W. Application and verification of different turbulence models in eddy simulation of axial flow pump tip leakage. *Chin. J. Agric. Eng.* **2013**, *13*, 46–53. (In Chinese) [CrossRef]

MDPI
St. Alban-Anlage 66
4052 Basel
Switzerland
Tel. +41 61 683 77 34
Fax +41 61 302 89 18
www.mdpi.com

Water Editorial Office
E-mail: water@mdpi.com
www.mdpi.com/journal/water

www.ingramcontent.com/pod-product-compliance
Lightning Source LLC
LaVergne TN
LVHW070908170726
843515LV00005B/732

* 9 7 8 3 0 3 6 5 5 8 5 8 5 *